ANNALS *of* THE NEW YORK ACADEMY OF SCIENCES

EDITOR-IN-CHIEF
Douglas Braaten

ASSOCIATE EDITOR
Rebecca E. Cooney

PROJECT MANAGER
Steven E. Bohall

Artwork and design by Ash Ayman Shairzay

T0344847

The New York Academy of Sciences
7 World Trade Center
250 Greenwich Street, 40th Floor
New York, NY 10007-2157

annals@nyas.org
www.nyas.org/annals

**The New York
Academy of Sciences**

Published by Blackwell Publishing
On behalf of the New York Academy of Sciences

Boston, Massachusetts
2012

ANNALS *of* THE NEW YORK ACADEMY OF SCIENCES

VOLUME
1267

ISSUE

Effects of Genome Structure and Sequence on Variation and Evolution

ISSUE EDITOR

Lynn H. Caporale

TABLE OF CONTENTS

1 Overview of the creative genome: effects of genome structure and sequence on the generation of variation and evolution
Lynn Helena Caporale

11 Genome hyperevolution and the success of a parasite
J. David Barry, James P. J. Hall, and Lindsey Plenderleith

18 The tricky path to recombining X and Y chromosomes in meiosis
Liisa Kauppi, Maria Jasin, and Scott Keeney

24 Sites of genetic instability in mitosis and cancer
Anne M. Casper, Danielle M. Rosen, and Kaveri D. Rajula

31 The genome: an isochore ensemble and its evolution
Giorgio Bernardi

35 Multiple levels of meaning in DNA sequences, and one more
Edward N. Trifonov, Zeev Volkovich, and Zakharia M. Frenkel

39 Evolution of simple sequence repeat–mediated phase variation in bacterial genomes
Christopher D. Bayliss and Michael E. Palmer

45 Indirect selection of implicit mutation protocols
David G. King

53 G4 motifs in human genes
Nancy Maizels

61   Adaptive radiation of venomous marine snail lineages and the accelerated evolution of venom peptide genes
Baldomero M. Olivera, Maren Watkins, Pradip Bandyopadhyay, Julita S. Imperial, Edgar P. Heimer de la Cotera, Manuel B. Aguilar, Estuardo López Vera, Gisela P. Concepcion, and Arturo Lluisma

71   Integrons and gene cassettes: hotspots of diversity in bacterial genomes
Ruth M. Hall

79   Creative deaminases, self-inflicted damage, and genome evolution
Silvestro G. Conticello

86   Three-dimensional architecture of the IgH locus facilitates class switch recombination
Amy L. Kenter, Scott Feldman, Robert Wuerffel, Ikbel Achour, Lili Wang, and Satyendra Kumar

95   Preaching about the converted: how meiotic gene conversion influences genomic diversity
Francesca Cole, Scott Keeney, and Maria Jasin

103   Gross chromosomal rearrangement mediated by DNA replication in stressed cells: evidence from *Escherichia coli*
J.M. Moore, Hallie Wimberly, P.C. Thornton, Susan M. Rosenberg, and P.J. Hastings

110   Implications of genetic heterogeneity in cancer
Michael W. Schmitt, Marc J. Prindle, and Lawrence A. Loeb

---

117   Corrigendum for Ann. N.Y. Acad. Sci. 2009. 1182: 47–57

Ann. N.Y. Acad. Sci. ISSN 0077-8923

ANNALS OF THE NEW YORK ACADEMY OF SCIENCES

Issue: *Effects of Genome Structure and Sequence on Variation and Evolution*

Overview of the creative genome: effects of genome structure and sequence on the generation of variation and evolution

Lynn Helena Caporale

Address for correspondence: caporale@usa.net

This overview of a special issue of *Annals of the New York Academy of Sciences* discusses uneven distribution of distinct types of variation across the genome, the dependence of specific types of variation upon distinct classes of DNA sequences and/or the induction of specific proteins, the circumstances in which distinct variation-generating systems are activated, and the implications of this work for our understanding of evolution and of cancer. Also discussed is the value of non text-based computational methods for analyzing information carried by DNA, early insights into organizational frameworks that affect genome behavior, and implications of this work for comparative genomics.

Keywords: Genome; evolution; DNA structure; genetic code; mutation; Darwin

"A grand and almost untrodden field of inquiry will be opened, on the causes and laws of variation"

—Charles Darwin[1]

Introduction

Our understanding of genome organization and hierarchies of control are in their infancy, yet we are conversant with terms such as "locus control region,"[2] "allelic exclusion,"[3,4] "boundaries," "insulators," and "imprinting."[5] It no longer is assumed that a DNA sequence functions identically wherever a researcher places it in the genome.[6,7] Some regions are gene rich, some are gene "deserts"[7] that burst with RNA transcripts,[8] while the creation of protein-coding sequences from cryptogenes[9] makes it clear that there is much more for us to understand about recognition, integration, and storage of information in genomes.

While most attention is focused on biochemical regulation during the life of an individual, much of the focus in this issue of *Annals of the New York Academy of Sciences*[a] is on the organization and regulation of mechanisms that effect genetic change and thus affect the survival of lineages of organisms over evolutionary timescales. This discussion points to the importance of nontraditional approaches to the analysis of genome sequences and has implications ranging from where to place genes in the genome and what sequences to use for genetic engineering, to combating pathogens and tumors, to evolutionary theory. While this overview cannot be a comprehensive review of such a broad range of subjects, it is meant to highlight common themes among the papers in this issue, and to point the reader to the individual contributions for more in-depth discussion.

Much as a genome serves as a framework in which new information arises and operates, the theory

[a]"Effects of Genome Structure and Sequence on the Generation of Variation and Evolution." Ed. Lynn H. Caporale. Special issue of *Ann. N.Y. Acad. Sci.* (2012) 1267, 1–10. The publication and this overview derive from an interdisciplinary international conference "Effects of Genome Structure and Sequence on the Generation of Variation and Evolution" hosted by The Center for Discrete Mathematics and Theoretical Computer Science (DIMACS).[10]

doi: 10.1111/j.1749-6632.2012.06749.x
Ann. N.Y. Acad. Sci. 1267 (2012) 1–10 © 2012 New York Academy of Sciences.

of evolution has served as a framework for our deepening understanding of biology. When Darwin[1] and Wallace[11] proposed that evolution results from the action of selection on heritable variation, they did not know how a trait could be inherited. Genes, and mutations, were incorporated into evolutionary theory during the first half of the 20th century,[12] before the biochemical basis of mutation was understood and before there was any concept of how DNA encodes information—in fact, before it was known that DNA is the genetic material. Updated to include DNA but before a single genome had been sequenced,[13] evolutionary theory, as currently taught,[12] assumes that all mutations are "random[b]." But the statement that all mutations are random was not Darwin's idea.

> "I have ..sometimes spoken as if the variations . . . had been due to chance.
> This, of course, is a wholly incorrect expression, but it serves to acknowledge plainly our ignorance of the cause of each particular variation."[14]

According to the theory of evolution, selection acts upon variation; thus selection should act upon DNA sequence–dependent variations in the probability of distinct classes of mutation. In other words, in contrast to what is widely taught, evolutionary theory actually predicts that mutation will *not* always be random, and, as described by multiple papers in this issue, the probability of distinct classes of genetic variation is, in fact, not equal everywhere in a genome.

Genome organization and genetic variation

Regional differences in distinct types of variation

Barry *et al*. describe extreme regional differences in the generation of variation in the trypanosome

---

[b]It is important to clarify the meaning of "random mutation," as two meanings often are confused: randomness with respect to DNA sequence and randomness with respect to potential adaptive value. Clearly, mutations are not randomly distributed along a DNA sequence; current statements of evolutionary theory assume that mutations must be random with respect to their potential adaptive value, but it is argued in this issue (*Ann. N.Y. Acad. Sci.* (2012) **1267**, 1–116) and elsewhere[67] that since the probability of different classes of mutation along a DNA sequence varies, this will, like all variation, "feel" the pressure of natural selection so that not all mutations will be random with respect to their potential adaptive value.

genome.[3] Mutation in much of the genome is repaired, but antigenic surface proteins (variable surface glycoproteins (VSGs)) located in subtelomeres diversify rapidly. [3]

Chromosomes have unique, cell type–dependent territories within eukaryotic nuclei that affect the probability of specific mutations.[15] As Kauppi *et al*. point out, both the position of chromosomes in three-dimensions of the nucleus and higher-order chromatin structure determine access to meiotic recombination machinery.[16]

Timing of replication and genome organization of function

That distinct regions of the genome replicate at different times enables both epigenetic mechanisms of control and regulated differences in the probability of distinct classes of mutation,[6] an area of investigation rich with potential for unexpected discoveries.

As described by Kauppi *et al*.,[16] distinct isoforms of SPO11 are induced early and late during meiotic recombination; pairing occurs later in the pseudoautosomal region of the X and Y chromosomes. Many cold spots of meiotic recombination replicate late during pre-meiotic replication (chromatin alterations in pre-meiotic replication influence the location of double-strand breaks in meiosis).[17] The timing of replication at imprinted loci depends on the parent of origin.[18] Casper *et al*.[19] note that some fragile sites may lie at transition between early and late replicating DNA, which also appear to be boundaries of isochores (described by Bernardi[7] as functionally distict blocks of differing G/C content in genomes).

In some ciliates up to 95% of the germline genome sequences are physically eliminated in an RNA-dependent process during the formation of the (somatic) macronucleus,[20,21] providing a window into RNA-dependent definition of special contexts or behaviors in a genome. Regions of DNA that are destined to be left out of the somatic nucleus of ciliates replicate during a second phase of DNA synthesis.

Arber[22] has asked us to reflect upon how different our perspective would be if we had chosen a ciliate as a model unicellular organism instead of *Escherichia coli*.

Creative DNA sequences

Slippery DNA

In a broad range of species[23,24] from bacteria to humans, repetitive DNA sequences, such as TTTT and

CAGCAGCAG, gain and lose units of the repeat at a rate that can be >1000 times the genome's average mutation rate, as discussed in by Trifonov et al.,[25] Bayliss and Palmer,[26] Moxon et al.,[27] and King.[28] Subpopulations with decreased levels of mismatch repair[29] further increase the rate of exploration of repeat lengths.[26]

Such repetitive sequences have been selected against in some regions of the genome, but it is important to note evidence that these mutagenic sequences apparently are selected *for* where the diversity that they generate is adaptive.[28] For example, these reversibly mutable sequences are enriched in "contingency genes,"[26,27] such as those encoding bacterial coat proteins, facilitating adaptation to a rapidly changing environment by creating a high level of what Barry et al. term "pre-emptive variation" in populations.[3]

DNA sequences create structures that attract genetic change

When Watson and Crick presented us with the double helical structure of DNA,[30] they lacked the experimentalist Rosalind Franklin's awareness of DNA's unevenness and of the sensitivity of DNA's structure to its environment;[31] their beautiful, simple model neither captures nor even suggests the full range of structures of DNA itself.[32,33]

Given the evenness of the iconic Watson-Crick structure, computational analyses have treated the bases as if they are actual letters, "A," "T," "G," and "C." Yet, DNA carries information in forms that are not limited to simple strings of text. For sequences highly enriched in one base or in which a sequence motif is repeated, the distinct tilt and twist of different steps of stacked base pairs along the helix cause significant variations in flexibility and substantial deviations from the iconic DNA structure;[33] further complexity is introduced as this structure interacts with RNA and proteins, and is acted upon by enzymes whose fidelity is affected by the varied structures.[34–36]

Structure-forming sequences often stall transcription and replication and cause recombinogenic double-strand breaks

Stable non-iconic DNA structures that tend to break in mitosis are sites of frequent chromosome breaks and rearrangements in tumor cells.[19]

Maizels[36] discusses G4 motifs ($G_{\geq 3}N_xG_{\geq 3}N_xG_{\geq 3}N_xG_{\geq 3}$), which underlie some of the most poly-

morphic sites in the human genome. The uneven and characteristic distribution of G4 motifs among genes and gene regulatory regions suggests evolutionary selection for and against G4 sequences, which form stable structures when DNA is single-stranded (i.e., during transcription, replication or recombination).

Repeats of $(TAA)_n$ in the genome of African trypanosomes also form structures that can stall DNA replication, resulting in double-strand breaks that generate diversity focused on VSGs, as discussed by Barry et al.[3] These breaks are repaired by recombination with a "palette" of hundreds of unexpressed VSG genes, resulting in switches in this antigenic coat at a rate of 10^{-3} /cell/generation, thus confronting host immunity with challenging trypanosome population diversity among, for example, the estimated 10^{11} individuals that infect each cow, and which can spread among cows.[3]

Frameworks for variation

Pseudogenes. As Barry et al. describe,[3] trypanosomes create further templated, mosaic diversity through gene conversion from an archive of nearly 2000 silent VSG genes and pseudogenes. Thus, descendants of a single trypanosome can generate a repertoire of protein sequences that is orders of magnitude larger than what is encoded explicitly in its DNA (much as the vertebrate immune system that trypanosomes battle also generates diversity from gene fragments along with rules for their assembly[37]).

As described in the contribution by Olivera et al.,[38] the genome of each cone snail species encodes hundreds of toxins. The extraordinary DNA sequence conservation in the toxin gene signal sequence (without even "silent" mutations) inspires the speculation that a conserved DNA framework may enable generation of the explosive variation observed in the C-terminal active toxin region.

Cassettes. Hall[39] describes bacterial cassettes, which, generally, contain single genes bordered by *attc* sites that are recognized by the integrases of integrons. Integrons can capture and release hundreds of gene cassettes and can act as expression vectors; incorporated into mobile elements, integrons provide a framework that enables information to flow horizontally among bacteria.

Implications for sequence analysis

Functionally important non-iconic structures are pervasive

As described by Trifinov et al.,[25] embedded in CG-GAAATTTCCG, the "nucleosome positioning chromatin code", is GAAAATTTTC, which, when periodically repeating, curves DNA and is a dominant motif across the genomes of prokaryotes as well as eukaryotes.

Because deviation from the standard B-form DNA coordinates can be propagated into neighboring sequences, analyzing the physical chemical properties of DNA sequences is not as straightforward as looking up codons in a table. In addition, during transcription, replication, and recombination DNA forms transient, complicated multi-chain structures that are sequence-dependent.[32–34,36,40]

Enzymatic recognition of RNA structures that form due to relationships among neighboring nucleotide sequences are highlighted in Conticello's discussion[41] of A > I editing deaminases, which alter codons, the splicing of mRNA, and the processing of miRNA. Some deaminases, such as ADARS, recognize double-stranded regions of RNA that form due to complementary sequences on the same strand. As we increasingly recognize the myriad roles of RNA—which forms complex structures that can be altered by metabolites[42]—the need to go beyond linear text-based analysis to extricate understanding of function from genome sequence becomes obvious and compelling.

Conservation of function is not synonymous with conservation of DNA sequence

Algorithms are needed that search for sequences that share categories of behavior, such as attracting mechanisms that generate focused variation.

Structure-forming sequences. Maizels[36] points out that while the vast majority of human exons contain no G4 motifs, in nearly half of human genes these motifs are found on the nontemplate strand at the 5′ end of the first intron. The ability of DNA to form structures is conserved at these sites from humans to amphibians, but the DNA sequences themselves are not conserved.

Hypervariable sequences. As King[28] points out, far from being recognizable by their sequence conservation, sequences important for rapid adaptation can be dismissed as unimportant due to their low conservation. However, some sites of repetitive DNA have been deeply conserved even when the nature of the repeat (i.e. the actual repeating sequence) was not conserved. What is conserved is not the sequence but the property of instability.[6]

There is no A in DNA

The need for non-text based DNA analysis is illustrated by the example of non-random insertion sites of the P transposon, for which no text-based consensus was identified; an analysis that treated DNA as a molecule with physical chemical properties, rather than simply as text, revealed that insertion sites were palindromes in the hydrogen-bond donor/acceptor pattern at the edges of base pairs as viewed from the major groove.[43]

Overlapping Information

Multiple levels of information can be carried through a single DNA sequence.[6,25,44] For example, the consensus sequence that is the focus of hypermutation in immunoglobulin V regions constrains the choice among otherwise synonymous codons.[41]

Biochemical analysis of evolutionary history

That a "mutation phenotype" is created by DNA sequences in the context of enzymes that act upon DNA is illustrated by mutations in the enzyme that initiates hypermutation, activation-induced deaminase (AID) that in turn alter the mutation pattern in antibody genes.[41] Due to the long-held assumption of the complete randomness of all mutations, analysis of genome evolution in light of biochemical mechanisms is an unexplored field. Yet, as Conticello[41] points out it is possible to observe footprints of the past action of specific enzymes on genomic DNA.

Regulated mutations

Generation of diversity through meiosis

In considering whether genomes might have evolved biochemical mechanisms that actively generate variation, King suggests[28] we reflect upon the biochemical investment in meiosis, during which human chromosomes assort into haploid gametes with 2^{23} possible combinations. As described in the contribution by Cole et al.[45] the diversity of the assorted chromosomes is further increased when SPO11 creates >200 double-stranded breaks in each cell that are "repaired" using not the sister chromatid but the non-identical homolog (which comes from the other parent). By using a highly polymorphic mouse hotspot Cole and her coworkers are able to detect a majority of noncrossover events at that site and to measure the extent of greater than Mendelian transmission of those alleles used to repair, via gene conversion, the breaks made in their homolog.

The contribution of both local sequence and protein context to the generation of genome variation is illustrated by the role of a meiosis-specific histone H3 methyltransferase, with rapidly changing target sites, in the location of SPO11-dependent breaks. Kauppi et al. investigate the mechanism underlying the observation that the 0.7 Mb pseudo autosomal region of spermatocytes has a double strand break frequency that is 10- to 20-fold higher than the genome average in meiosis, emphasizing in the discussion that to understand evolution in eukaryotes we must pay attention to what happens in the germline.[16]

Inducible focused genetic change: some examples from the immune system

Protocols. What is inherited is not intact immunoglobulin genes,[46] but rather protocols[37] that underlie the combinatorial assembly of immunoglobulin variable regions from a palette of germline gene fragments, generating a diversity of antigen combining sites orders of magnitude larger than could be encoded explicitly in the genome.[6] Generation of immune system diversity by novel mechanisms is not limited to vertebrates.[47]

Induced region-specific recombination. When confronted with an antigen, regionally targeted DNA breaks can switch the class of the previously assembled heavy chain variable region to an effector function that is encoded by one of a selection of constant (C_H) region genes (a process called class switch recombination). Notably, the choice of effector and thus the switch (S) region targeted for recombination is regulated by the environment (i.e., it depends upon the type of antigen). The contribution by Kenter et al.[2] takes us through the steps centered on transcription, initiated upstream of the C_H region that is targeted for recombination and which is paused by a structure-forming region.[36] Multiple levels of control, embodied in biochemical mechanisms that act across > 100 kb, juxtapose the two S regions targeted for the class switch. Recombination then is initiated by deamination by AID (Conticello[41] observes that AID's creation of tar-

geted DNA damage, which both triggers class switch and focuses somatic diversification of immunoglobulin variable regions, predates the vertebrate immune system, which suggests as yet undiscovered roles.)

Mutations from genetic insult depend on the balance of expressed proteins. Simply decreasing the expression of repair proteins can increase mutation dramatically, for example at Cs, which spontaneously deaminate to U.[48] The immune system accelerates and focuses C deamination through the induction of AID, which enzymatically changes Cs to Us in DNA encoding immunoglobulin variable regions.[41] Whether deamination leads to variation via mutagenic replication across the gap or by gene conversion is determined by the balance among proteins that define distinct pathways of repair, such as translesion synthesis or recombination.[49,50]

Stress
As McClintock first proposed,[51] stress can trigger genome reorganization in plants through the release of transposable elements.[52] In bacteria, maladaptation to the environment, i.e. stress (for example, starvation), increases mutation and thus genetic diversity in a manner that is dependent upon specific gene products.[53–58] Stress also increases gene sharing among bacteria, for example, by inducing prophage[53] and integron integrases.[39] As Evelyn Witkin observed,[59] "we used to think the mutagen went ZAP! and that was that." But now we understand that the outcome of a potentially mutagenic insult is affected by the balance of myriad proteins in a cell.[48,60]

As we gain the ability to parse genomic sequences into regulatory networks, we will better understand how the stress of challenging environments is sensed, and the ways in which downstream responses affect genome variation.[61,62] Computational analyses by Benham and Wang suggests that changes in DNA supercoiling in response to environmental or physiological stresses could affect transcription genome-wide at specific classes of sites.[63]

Hastings et al.[56] observe that distinct proteins are required for mutations in growing cells, for frameshift (point) mutations in cells under nutrient stress, and for gene amplification in stressed cells. Under stress, some cells become permissive for using nearby single-stranded DNA of even only very short homology to restart collapsed replication forks, creating a duplication if this "micro" homologous sequence already has replicated (a deletion if this occurs downstream of the collapse). Such chromosomal changes are a risky source of genomic creativity; only a sub-population of cells runs the risky experiment.[56]

Implications for understanding evolving systems

Second-order selection and the survival of lineages

As King points out,[28] laboratory-created mutations may be predominantly deleterious, but for mutations arising spontaneously under natural conditions (with their own mix of "synonymous" codons and in the sequence context in which they evolved[6]) the ratio of benefit to harm has not been defined.

The assumption that most naturally occurring mutations are deleterious random errors is challenged by an extensive and expanding literature. Significant standing variation[28] is generated by reversible changes in the number of units in repeating DNA sequences, by transposable elements, site-directed recombination, and meiotic recombination and results in an unevenness of the probability of distinct classes of mutation across a genome. The spread of traits, such as antibiotic resistance by integrons,[39] also generates adaptive diversity rapidly. That mutations at repetitive sequences are reversible ensures that an adaptation is retained by a lineage, enabling the population of descendants of an individual bacterium to express a mixture of traits and thus be adapted to survive in a broader range of recurring conditions than any individual parent.

In adapting to new and challenging environments, bacteria with elevated mutation rates ("mutators") out-compete the wild-type cells.[64,65] As pointed out by Bayliss and Palmer,[26] when bacteria move within and between hosts they repeatedly experience a plunge in population size and thus lose genetic diversity just at the moment when they need it most. Thus, as discussed by King,[28] the assumption that selection acts only to lower mutation rates makes sense only for well-adapted organisms in a stable environment.[6]

Under challenging circumstances, mechanisms that focus diversity precisely where past selection has proven diversity to be adaptive can provide a dramatic survival advantage. For example, when highly specific antibodies directed against a major surface antigen appear, individuals that *already* express a different antigen survive. As pointed out by Barry *et al.*,[3] individual variants of trypanosome *VSG* are exposed to direct selection only when they actually are expressed, which is rare. However what can, and has been, selected, are the multiple biochemical mechanisms that generate the VSG diversity that challenges host immunity. Barry *et al.* state, "the view that evolution acts on only the immediate phenotype, rather than on long-term strategy is not compatible with what is observed for trypanosomes."

Olivera *et al.*[38] track the pattern of divergence in a large gene superfamily where the phylogenetic relationships among species are defined. In discussing the "startlingly rapid" evolution of cone snail venom peptides he observes that genes important for interactions with other organisms, which he names "exogenes," have high rates of evolution. This point is made also by Conticello[41] in discussing the dramatic expansion in primates of the APOBEC3 deaminases, which are involved in restriction of retroviruses and mobile elements, and resonates with the discussion of contingency genes and pre-emptive variation.

Thus, various examples described above, from immunoglobulins to bacteria to trypanosomes to cone snails, suggest that generation of genome variation is not limited to mutations that are both "rare and random;" and while these all are examples of pathogen/immunity and predator/prey challenges,[66] it increasingly appears unlikely that such focused variation is limited to such examples. As King[28] points out, variation generators such as simple sequence repeats and transposable elements affect a large proportion of functional genetic loci.

In fact, identifying the genomic locations at which high levels of variation occur may suggest what characteristics of a lineage benefit from expedited exploration, providing a new window into the challenges faced by that lineage of organisms.[67] For example, a set of genetic adaptations turns up repeatedly in the evolution of distinct species of freshwater-dwelling sticklebacks from marine sticklebacks.[68] Interestingly, in comparing fish that live in different environments, there are repeated inversions at the same chromosomal locations, reminiscent of reversible inversions in bacteria,[60] such as an inversion that leads to the synthesis of bacterial fimbrae, an inversion that is favored at temperatures that occur inside the host, while the orientation that does not express the gene is favored outside the host.[69]

It is reasonable to consider that natural selection would act indirectly on the mechanisms that generate genome variation much as it acts directly on beaks and wings. As Darwin suggested, "Why...should nature fail in selecting [useful] variations...? I can see no limit to this power."[1]

Tumors

Mutations that decrease repair and increase mutations increase the risk of cancer.[70] Much as a mutator phenotype facilitates rapid adaptation by generating diverse bacteria,[65] Loeb and colleagues suggest[64] that such a genome-wide elevation in the rate of mutation enables once normal cells to overcome redundant restraints on division, and to gain the ability to invade, to metastasize, and to resist treatment. Even given the constraints of current sequencing technology, Loeb identified over two orders of magnitude more mutations in a tumor than in paired adjacent normal tissues. Such an enormous reservoir of genetic diversity recalls the "pre-emptive diversity" at bacterial contingency loci[26,27] and trypanosome VSGs.[3] Thus, tumors can be viewed as a mix of evolving genomes.[64,71,72]

Simply altered levels of expression of unmutated enzymes, including helicases that unwind G4- structures[36] and AID and other deaminases,[41] can contribute to tumor initiation and progression, as can increased expression of trans-lesion polymerases, release of transposons,[73] and alterations in the balance among unmutated proteins involved in a wide range of biochemistry, from nucleotide pools to repair.[60] The articles in this volume[c] suggest many mechanisms to expand our understanding of the generation of distinct classes of genome variation in tumors.

As Casper describes, simply decreasing the expression of *unmutated* polymerase alpha can lead to breaks at "fragile" sites where replication is stalled by DNA sequences that tend to form secondary structures when single stranded; more than half of the deletions, amplifications, and translocations observed in cancer have at least one breakpoint at a common fragile site (CFS).[19] Casper further finds that the loss of heterozygosity often observed in

cancer cells may result from repair of breaks by mitotic recombination between *homologous* chromosomes. Current cancer treatments, involving replication stress and hypoxic conditions, may increase the frequency of both events, further driving cancer progression. Hastings *et al.*[56] also discuss gross chromosomal rearrangements that occur during activated stress responses. Loeb and colleagues,[64] however, suggest that given a tumor with a mutator phenotype, further raising the mutation rate (as many chemotherapeutic agents and radiation do) might create a mutation crisis that destroys tumor cells. Thus, better understanding the biochemical mechanisms that generate the variation underlying the evolution of a tumor would have significant practical consequences for developing more effective strategies to treat each cancer. Further analysis of genetic alteration in tumors is facilitated by http://cbioportal.org.[74]

Summary

Taken together, the papers described above work to expand our imagination as to the types of information embedded in genomes, and how that information can be organized, integrated, and represented; increases both in the sensitivity of nucleic acid sequencing and in computational power will provide the capability to ask the deeper questions that this inquiry inspires.

Nontraditional approaches to the analysis of genome sequences, such as algorithms that treat DNA sequences as something other than text, are needed to recognize specific classes of structure and to define new contexts and classes of functions (protocols) across the genome. With the burst of discoveries of novel classes of RNA, the need for nontext base analysis has become glaringly obvious. As we step back from the evenness of the iconic Watson-Crick structure it becomes apparent that non-iconic structures are pervasive and functionally important.

Recognizing that conservation of function is not synonymous with conservation of DNA sequence, algorithms are needed that search for sequences that share categories of behavior; distinct sequences may create the same class of structure or share other properties (e.g. mutability). A single G can create part of a codon, of a hypermutable sequence, a secondary structure, or all simultaneously. Selection

---

[c]*Ann. N.Y. Acad. Sci.* (2012) 1267, 1–116.

acts on the multiple levels of information that are carried in DNA sequences.

Much of the focus of the papers in this issue is on the organization and regulation of mechanisms that generate variation. As illustrated by repetitive sequences that undergo repeated reversible mutations, mutation is not random along a DNA sequence. There are regional differences in distinct types of variation, some enabled by frameworks that facilitate distinct forms of exploration, such as pseudogenes and cassettes. We find "pre-emptive diversity" in bacterial contingency loci, trypanosome variable surface glycoproteins, and in effect, in tumors. DNA sequencing sensitivity is reaching the point where we can sequence and analyze the actual diversity generated in each individual in a population of bacteria and individual haploid human gametes, and in evolving tumor cells, rather than simply infer diversity-generating mechanisms by observing the results of selection.

Identifying genomic locations at which significant standing variation occurs in a lineage can point us to a biological role for expedited exploration of specific traits as the lineage confronts challenges that are repeated in its environment. As illustrated dramatically by the vertebrate immune system, both local sequence and the regulated protein context contribute to the generation of distinct classes of variation. Stress (e.g. starvation), can be sensed biochemically; the outcome of a potentially mutagenic insult are affected by the balance of myriad proteins expressed in a cell.

The power of a better understanding of genomic mechanisms that actively generate variation ranges from the practical to the theoretical. The most immediate practical applications are likely to be identification of novel targets to combat pathogen defenses and the development of a more strategic and effective arsenal against tumors. The research described above investigates multiple levels of genome organization and architecture, and evolution. The author points the reader to the individual contributions in this volume for more in depth discussion.

Conflicts of interest

The author declares no conflicts of interest.

References

1. Darwin, C. *On the Origin of Species*. Chapter 14: Recapitulation and Conclusions.
2. Kenter, A., S. Feldman, R. Wuerffel, *et al.* 2012. Three-dimensional architecture of the Igh locus facilitates class switch recombination. *Ann. N.Y. Acad. Sci.* **1267:** 86–94. This volume
3. Barry, J.D., J.P.T. Hall & L. Plenderleith. 2012. Genome hyperevolution and the success of a parasite. *Ann. N.Y. Acad. Sci.* **1267:** 11–17. This volume
4. Vettermann, C & M.S. Schlissel. 2010. Allelic exclusion of immunoglobulin genes: models and mechanisms. *Immunol. Rev.* **237:** 22–42.
5. Murrell, A. 2011. Setting up and maintaining differential insulators and boundaries for genomic imprinting. *Biochem Cell Biol.* **89:** 469–478.
6. Caporale, L.H. 2006. An Overview of The Implicit Genome. In *The Implicit Genome*. Ed., L. Caporale: Oxford University Press.
7. Bernardi, G. 2012. The genome: an isochore ensemble and its evolution. *Ann. N.Y. Acad. Sci.* **1267:** 31–34. This volume
8. Mattick, J.S. 2009. The Genetic Signatures of Noncoding RNAs. *PLoS Genet.* **5**(4): e1000459.
9. Aphasizhev, R. & I. Aphasizheva. 2011. Uridine insertion/deletion editing in trypanosomes: a playground for RNA-guided information transfer. *Wiley Interdiscip. Rev. RNA* **2:** 669–85.
10. http://www.adaptivegenome.net/
11. Wallace, A.R. 1991 [1869]. *The Malay Archipelago*. Oxford University Press. p. 419.
12. Barton, N.H., D.E.G. Briggs, J.A. Eisen, *et al.* 2007. *Evolution.* Cold Spring Harbor Laboratory Press, Cold Spring Harbor, N.Y.
13. Sanger, F., G.M. Air, B.G. Barrell, *et al.* 1977. Nucleotide sequence of bacteriophage phi X174 DNA. *Nature* **265:** 687–695.
14. Darwin, *op. cit.* Chapter 5: Laws of Variation.
15. Baker, M. 2011. Genomes in three dimensions. *Nature* **470:** 289–294.
16. Kauppi, L., M. Jasin, & S. Keeney. 2012. The tricky path to recombining X and Y chromosomes in meiosis. *Ann. N.Y. Acad. Sci.* **1267:** 18–23. This volume
17. Borts, R. & D.T. Kirkpatrick. 2006. The role of the genome in the initiation of meiotic recombination. In *The Implicit Genome*. Ed., L. Caporale: Oxford University Press.
18. Shufaro, Y., O. Lacham-Kaplan, B.-Z. Tzuberi, *et al.* 2010. Reprogramming of DNA Replication Timing. *Stem Cells* **28:** 443–449.
19. Casper, A., D. Rosen & K. Rajula. 2012. Sites of genetic instability in mitosis and cancer. *Ann. N.Y. Acad. Sci.* **1267:** 24–30. This volume
20. Jahn, C.L. 2006. Nuclear Duality and the Genesis of Unusual Genomes in Ciliated Protozoa. In *The Implicit Genome*. Ed., L. Caporale: Oxford University Press.
21. Nowacki, M., K. Shetty & L.F. Landweber. 2011. RNA-Mediated Epigenetic Programming of Genome Rearrangements. *Annu. Rev. Genomics Hum. Genet.* **12:** 367–89.
22. Arber, W. 1999. Concluding Remarks. *Ann. N.Y. Acad. Sci.* **870:** 344–345.
23. Fondon, J.W. III & H.R. Garner. 2007. Detection of length-dependent effects of tandem repeat alleles by 3-D geometric decomposition of craniofacial variation. *Dev. Genes Evol.* **217:** 79–85.

24. Fondon, J.W. III, A. Martin, S. Richards, *et al.* 2012. Analysis of Microsatellite Variation in *Drosophila melanogaster* with Population-Scale Genome Sequencing. *PLoS One* 7(3): e33036.

25. Trifonov, E.N., Z. Volkovich & Z.M. Frenkel. 2012. Multiple levels of meaning in DNA sequences, and one more. *Ann. N.Y. Acad. Sci.* **1267**: 35–38. This volume

26. Bayliss, C.D. & M.E. Palmer. 2012. Evolution of simple sequence repeat-mediated phase variation in bacterial genomes. *Ann. N.Y. Acad. Sci.* **1267**: 39–44. This volume

27. Moxon, R., C. Bayliss & D. Hood. 2006. Bacterial contingency loci: the role of simple sequence DNA repeats in bacterial adaptation. *Annu. Rev. Genet.* **40**: 307–33.

28. King, D.G. 2012. Indirect selection of implicit mutation protocols. *Ann. N.Y. Acad. Sci.* **1267**: 45–52. This volume.

29. Rocha, E.P.C. 2006. The role of repeat sequences in bacterial genetic adaptation to stress. In *The Implicit Genome*. L. Caporale Ed.: Oxford University Press.

30. Watson, J.D. & F.H. Crick. 1953. Molecular structure of nucleic acids; a structure for deoxyribose nucleic acid. *Nature* **171**: 737–738.

31. Klug, A. 1974. Rosalind Franklin and the Double Helix. *Nature* **243**: 787–788.

32. Sinden, R.S., V.N. Potaman, E.A. Oussatcheva, *et al.* 2002. Triplet repeat DNA structures and human genetic disease: dynamic mutations from dynamic DNA. *J Biosci. Suppl.* **1**: 53–65.

33. Zheng, G., A.V. Colasanti, X.-J. Lu, & W.K. Olson. 2010. *DNALandscapes*: a database for exploring the conformational features of DNA. *Nucleic Acids Res.* **38**: D267–D274.

34. Shchyolkina, A.K., D.N. Kaluzhny, D.J. Arndt-Jovin, *et al.* 2006. Recombination R-triplex: H-bonds contribution to stability as revealed with minor base substitutions for adenine. *Nucleic Acids Res.* **34**: 3239–3245.

35. Kunkel, T.A. 2011. Balancing eukaryotic replication asymmetry with replication fidelity. *Curr. Opin. Chem. Biol.* **5**: 620–626.

36. Maizels, N. 2012. G4 motifs in human genes. *Ann. N.Y. Acad. Sci.* **1267**: 53–60. This volume.

37. Doyle, J., M. Csete & L.H. Caporale. 2006. An Engineering Perspective: The Implicit Protocols. In *The Implicit Genome*. Ed., L. Caporale: Oxford University Press.

38. Olivera, B.M., M. Watkins, P. Bandyopadhyay, *et al.* 2012. Adaptive radiation of venomous marine snail lineages and the accelerated evolution of venom peptide genes. *Ann. N.Y. Acad. Sci.* **1267**: 61–70. This volume.

39. Hall, R.M. 2012. Integrons and gene cassettes: hotspots of diversity in bacterial genomes. *Ann. N.Y. Acad. Sci.* **1267**: 71–78. This volume.

40. Kuduvalli, P.N., J.E. Rao & N.L. Craig. 2001. Target DNA structure plays a critical role in Tn7 transposition. *EMBO J.* **20**(4): 924–932.

41. Conticello, S.G. 2012. Creative deaminases, self-inflicted damage, and genome evolution. *Ann. N.Y. Acad. Sci.* **1267**: 79–85. This volume.

42. Ames, T.D., D.A. Rodionov, Z. Weinberg & R.R. Breaker. 2010. A eubacterial riboswitch class that senses the coenzyme tetrahydrofolate. *Chem. Biol.* **17**: 681–685.

43. Liao, G.C., E.J. Rehm & G.M. Rubin. 2000. Insertion site preferences of the P transposable element in Drosophila melanogaster. *Proc. Natl. Acad. Sci. USA* **97**: 3347–3351.

44. Caporale, L.H. 1984. Is there a higher level genetic code that directs evolution? *Mol. Cell. Biochem.* **64**: 5–13.

45. Cole, F., S. Keeney & M. Jasin. 2012. Preaching about the converted: How meiotic gene conversion influences genomic diversity. *Ann. N.Y. Acad. Sci.* **1267**: 95–102. This volume.

46. Hsu, E., N. Pulham, L.L. Rumfelt & M.F. Flajnik. 2006. The plasticity of immunoglobulin gene systems in evolution. *Immunol. Rev.* **210**: 8–26.

47. Terwilliger, D.P., K.M. Buckley, V. Brockton, *et al.* 2007. Distinctive expression patterns of 185/333 genes in the purple sea urchin, *Strongylocentrotus purpuratus*: an unexpectedly diverse family of transcripts in response to LPS, β-1,3-glucan, and dsRNA. *BMC Molecular Biology* **8**: 16.

48. Friedberg, E.C., G.C. Walker, W. Siede, *et al.* 2009. *DNA Repair and Mutagenesis*, 2nd Edition. ASM Press.

49. Neuberger, M.S. 2008. Antibody diversification by somatic mutation: from Burnet onwards. *Immunol. Cell Biol.* **86**: 124–32.

50. Beale, R. & Dagmar Iber. 2006. Somatic evolution of antibody genes. In *The Implicit Genome*. L. Caporale Ed.: Oxford University Press.

51. McClintock, B. 1984. The significance of responses of the genome to challenge. *Science* **226**: 792–801.

52. Naito, K., F. Zhang, T. Tsukiyama, *et al.* 2009. Unexpected consequences of a sudden and massive transposon amplification on rice gene expression. *Nature* **461**: 1130–1134.

53. Witkin, E.M. 1967. The radiation sensitivity of *Escherichia coli* B: a hypothesis relating filament formation and prophage induction. *Proc. Natl. Acad. Sci. USA* **57**: 1275–1279.

54. Radman, M. 1975. SOS repair hypothesis: phenomenology of an inducible DNA repair which is accompanied by mutagenesis. *Basic Life Sci.* **5A**: 355–367.

55. Foti, J.J., B. Devadoss, J.A. Winkler, *et al.* 2012. Oxidation of the guanine nucleotide pool underlies cell death by bactericidal antibiotics. *Science* **336**: 315–319.

56. Moore J.M., H. Wimberly, P.C. Thornton, *et al.* 2012. Gross chromosomal rearrangement mediated by DNA replication in stressed cells: evidence from *Escherichia coli*. *Ann. N.Y. Acad. Sci.* **1267**: 103–109. This volume

57. Storvik, K.A. & P.L. Foster. 2010. RpoS, the stress response sigma factor, plays a dual role in the regulation of Escherichia coli's error-prone DNA polymerase IV. *J. Bacteriol.* **192**: 3639–3644.

58. Shee, C., J.L. Gibson, M.C. Darrow, *et al.* 2011. Impact of a stress-inducible switch to mutagenic repair of DNA breaks on mutation in *Escherichia coli*. *Proc. Natl. Acad. Sci. USA* **108**: 13659–13664.

59. Friedberg, E.C. 1997. *Correcting the Blueprint of Life, an Historical Account of the Discovery of DNA Repair Mechanisms*. CSH Press. Cold Spring Harbor, N.Y. pp. 12–13.

60. Caporale, L.H. 2003. Natural Selection and the Emergence of a Mutation Phenotype. *Ann. Rev. Microbiol.* **57**: 467–485.

61. Singh, A.H., D.M. Wolf, P. Wang & A.P. Arkin. 2008. Modularity of stress response evolution. *Proc. Natl. Acad. Sci. USA* **105**: 7500–7505.

62. Fry, R.C., T.J. Begley & L.D. Samson. 2005. Genome-wide responses to DNA-damaging agents. *Annu. Rev. Microbiol.* **59:** 357–377.

63. Wang, H.Q. & C.J. Benham. 2008. Superhelical Destabilization in Regulatory Regions of Stress Response Genes. *PLoS Comp. Biol.* **4:** e17.

64. Schmitt, M., M. Prindle & L. Loeb. 2012. Implications of genetic heterogeneity in cancer. *Ann. N.Y. Acad. Sci.* **1267:** 110–116. This volume

65. Taddei, F., I. Matic, B. Godelle, & M. Radman. 1997. To be a mutator, or how pathogenic and commensal bacteria can evolve rapidly. *Trends Microbiol.* **5:** 427–429.

66. Wilkins, A.S. 2004. Diversity-generating mechanisms and evolution. *BioEssays* **27:** 111–112.

67. Caporale, L.H. 2005. Darwin in the genome. *BioEssays* **27:** 984.

68. Jones, F.C., M.G. Grabherr, Y.F. Chan, *et al.* 2012. The genomic basis of adaptive evolution in threespine sticklebacks. *Nature* **484:** 55–61.

69. Gally, D.L., J.A. Bogan, B.I. Eisenstein & I.C. Blomfield. 1993. Environmental Regulation of the fim Switch Controlling Type 1 Fimbrial Phase Variation in Escherichia coli K-12: Effects of Temperature and Media. *J. Bacteriol.* **175:** 6186–6193.

70. Friedberg, E.C., *et al. op cit*. Part 5 pp 863–1080 ISBN: 978-1-55581-319-2 ASM Press.

71. Caporale, L.H. 2002. Strategies as Targets, Round Two: Cancer. In *Darwin in the Genome*. Chapter 11, pp. 109–118. McGraw-Hill. New York.

72. Nik-Zainal, S., P. Van Loo, D.C. Wedge, *et al.* 2012. The life history of 21 breast cancers. *Cell* **149:** 994–1007.

73. Lee, E., R. Iskow, L. Yang, *et al.* 2012. Landscape of Somatic Retrotransposition in Human Cancers. *Science Express* (Jun 28) DOI:10.1126/science.1222077.

74. Cerami, E., J. Gao, U. Dogrusoz, *et al.* 2012. The cBio Cancer Genomics Portal: An Open Platform for Exploring Multidimensional Cancer Genomics Data. *Cancer Discov.* **2:** 401–404.

Ann. N.Y. Acad. Sci. ISSN 0077-8923

ANNALS OF THE NEW YORK ACADEMY OF SCIENCES

Issue: *Effects of Genome Structure and Sequence on Variation and Evolution*

Genome hyperevolution and the success of a parasite

J. David Barry, James P. J. Hall, and Lindsey Plenderleith

Wellcome Trust Centre for Molecular Parasitology, Institute of Infection, Immunity and Inflammation, College of Medical, Veterinary, & Life Sciences, University of Glasgow, Glasgow, United Kingdom

Address for correspondence: J. David Barry, Wellcome Trust Centre for Molecular Parasitology, Institute of Infection, Immunity & Inflammation, College of Medical, Veterinary & Life Sciences, University of Glasgow, 120 University Place, Glasgow G12 8TA, United Kingdom. dave.barry@glasgow.ac.uk

The strategy of antigenic variation is to present a constantly changing population phenotype that enhances parasite transmission, through evasion of immunity arising within, or existing between, host animals. Trypanosome antigenic variation occurs through spontaneous switching among members of a silent archive of many hundreds of variant surface glycoprotein (*VSG*) antigen genes. As with such contingency systems in other pathogens, switching appears to be triggered through inherently unstable DNA sequences. The archive occupies subtelomeres, a genome partition that promotes hypermutagenesis and, through telomere position effects, singular expression of *VSG*. Trypanosome antigenic variation is augmented greatly by the formation of mosaic genes from segments of pseudo-*VSG*, an example of implicit genetic information. Hypermutation occurs apparently evenly across the whole archive, without direct selection on individual *VSG*, demonstrating second-order selection of the underlying mechanisms. Coordination of antigenic variation, and thereby transmission, occurs through networking of trypanosome traits expressed at different scales from molecules to host populations.

Keywords: genome hyperevolution; antigenic variation; parasite; trypanosome; subtelomere

Introduction

Parasites and their hosts notoriously engage in an arms race, evolving measures and countermeasures against each other in a battle for supremacy. So strong are the selective forces, and consequently the rate and extent of evolution, that we can see a variety of extreme adaptations that are informative of mutational mechanisms and of general eukaryotic biology. Highly prominent in the arms race are phenotypes associated with evasion of mammalian immunity. Escape by a parasite might seem relatively simple—hide, or perhaps block, a key step in immunity, or maybe adopt camouflage—but closer inspection often reveals a highly complex biological system. The system often includes adaptations at various scales, ranging from the genome, through cellular structures and machinery, to population behavior within and between infections. Furthermore,

the functional outcomes of such adaptations generally are coordinated, forming a network of interacting processes that underpin a long-term strategy for enhanced transmission of the parasite and hence its persistence in an ecosystem. Nowhere is this complex type of evasion system more apparent than in the antigenic variation of African trypanosomes.

Antigenic variation: tactics for population survival

Antigenic variation is fundamental to the success of many pathogens. As infection proceeds, the resident pathogen population is decimated by populations of highly specific antibodies directed typically against its major surface antigen. Some individuals, however, have already switched to antigenically different versions of that molecule and survive and proliferate, a process that repeats many times, producing alternating waves of antigens and antibodies.[1] This process reflects the classical arms race in which pathogen survival mechanisms and host immune mechanisms coevolve. With the capacity to make $\sim 10^{10}$ distinct antibody idiotypes,

doi: 10.1111/j.1749-6632.2012.06654.x
Ann. N.Y. Acad. Sci. 1267 (2012) 11–17 © 2012 New York Academy of Sciences.

mammals seem destined to win this race. This enormous potential for immunological variation cannot be realized simply through direct encoding of antibodies in our genome, and requires a set of mechanisms that differentially shuffle and combine encoded components of antibodies.[2] How can a parasite, with a much smaller and less complex genome, compete?

In considering this question, we must broaden our vision. We have to adopt a systems biology approach, in which we link different functional scales of the parasite. Within infections, antigenic variation operates in concert with other intrinsic parasite parameters including growth and differentiation dynamics, to name just two.[3] Beyond individual infections, parasites have to be transmitted to new hosts. For a parasite that is abundant in the field, there is not an endless supply of new hosts, so a major challenge is likely to be successfully reinfecting hosts that already have antibodies corresponding to previous infections, either in circulation or stored as memory. Within an evolutionary framework, interactions at all scales, from molecules to host populations, amalgamate into a complex phenotype under complex selection. Features of antigenic variation, from molecules to parasite populations, will reflect the pressures of those selective forces.[3]

Trypanosomes live extracellularly in the blood, exposed to several immune mechanisms.[4,5] Their variant surface glycoprotein (VSG) forms a dense coat that thwarts innate immunity and prevents antibodies accessing invariant cell surface molecules and eliminating the infection.[6] A key event in the evolution of VSGs appears to have been dispensing with specific biochemical function, allowing their sequence to vary enormously, with merely some general structural and processing constraints. Spontaneously, at a rate of $\sim 10^{-3}$ switch/cell/generation, trypanosomes switch from one VSG to another, using mechanisms described below. The spontaneity and high rate of switching mean constant presence of variants before the onset of specific antibodies. For example, the first wave of parasites in a cow can peak with a total in excess of 10^{11} trypanosomes, and therefore $>10^8$ will have switched to a different variant, while a mouse can support $>10^8$ parasites with $>10^5$ emerging variants.[7] Many microbes facing strong challenges in hosts display such preemptive phenotypic variation, arising specifically from highly mutable contingency loci.[8]

In the trypanosome system, VSG switching is coordinated with growth and density-dependent differentiation of the parasite from proliferative to the nonproliferative transmission phase, which will infect tsetse flies on uptake in a bloodmeal.[9] There is a fairly constant supply of the transmission stage, the level of which is critically affected by the number of easily activated VSG variants. The more VSG variants that are present at a time, the less chance each variant subpopulation has of reaching the threshold to induce an antibody response, meaning that, when the number of variants is high, overall population numbers are predicted to be controlled by differentiation.[3,10] Such a swing toward differentiation-based control over immunity-based control results in higher levels of the transmission stage, but complete reliance on this mechanism would lead to prolonged high parasitemia and, therefore, earlier host death. In parallel, variants appear in ordered progression,[1] which favors efficient use of the archive and promotes chronicity of infection. Ordered expression could be achieved by variants clustering into "blocks" of distinct activation probability, with each block corresponding to a growth peak in the host.[3] As infections can run for years, many VSGs are probably required over the course of a single infection. Furthermore, it has been proposed that expressing substantially different subsets of VSGs would enable successful reinfection of previously exposed, partially immune hosts.[11] Reinfection would be facilitated also by the known diversity in the "metacyclic" VSGs expressed by the initial population injected by the tsetse fly vector.[12] On top of this reinfection requirement, the wide host range of trypanosomes introduces another dimension of uncertainty with which parasites must cope. Success in the wild, therefore, requires the various components of the network to adjust, enabling success in numerous host types.

Multiple mechanisms contribute to the generation of VSG diversity

The trypanosome genome appears to have evolved to provide such flexibility. To do so has required unusual adaptations, which operate at different levels.

Adaptations in genome structure

Expanded coding capacity and complexity. The enormous scale of antigenic variation is templated by an archive of nearly 2000 silent *VSG* genes and pseudogenes—nearly one-third of the core trypanosome genome.[13,14] The archive is very diverse, particularly in the sequences encoding the N-terminal domain of the VSG, which carries the key epitopes seen by the immune system. Typically, amino acid identity in that domain between VSGs is less than 20%, a particularly broad range that arises presumably from the lack of specific biochemical function and because many common amino acids specify alpha-helix, which comprises most of the domain.[6]

Partitioning of the genome: the importance of subtelomeres. As with many eukaryotes, the trypanosome genome is effectively partitioned into the core, containing genes under purifying selection and where most mutation is repaired, and subtelomeres, where multigene families associated with organismal phenotypic variation diversify rapidly.[15,16] The entire repertoire of *VSG*, including the archive and expressed genes, is located in subtelomeres.[14] Trypanosome subtelomeres are relatively enormous, in one case being several times longer than the associated chromosome core.[17]

Functionally, subtelomeres contribute two special features relevant to *VSG* and antigenic variation. The first of these features is ectopic recombination. Unlike chromosome cores, subtelomeres recombine ectopically, permitting sequence exchange or modification between different chromosomes, introducing diversity.[18] A number of recombination pathways can be involved.[19] Other, more general, mutational mechanisms contribute to *VSG* diversification, but ectopic recombination plays a particularly significant role and is likely to be specific to subtelomeres.

The second feature is the telomere position effect, in which gene promoters close to telomeres are subject to reversible repression mediated by silencing proteins that bind, directly or indirectly, to the telomere tract at the chromosome end. In the trypanosome, this effect appears to have evolved further into a sophisticated regulatory system. The only loci from which *VSG* can be transcribed are expression sites (ES), of which there are several in the genome.[20] All are telomere-proximal (i.e., the last genes before the telomere). Each ES comprises a promoter, several non-*VSG* genes, a set of imperfect repeats ("70-bp repeats"), and the *VSG* adjacent to the telomere. Antigen switching involves mostly

replacement, by gene conversion, of the expressed *VSG* copy (or a part thereof), by all (or part) of another *VSG*.[1,21] Only one ES is active at a time, which requires occupancy of a specific nuclear niche, the ES Body (see below), and inactivity of the other ES involves a telomere position effect.[22] It is likely that telomere proximity also functions in antigen switching, due to the capacity of telomeres to interact with one another, promoting recombination.[18]

Supernumerary chromosomes. The trypanosome has evolved a set of ∼100 nuclear minichromosomes, which carry, with little exception, only telomere-proximal *VSG* genes.[23] The two consequences are increase in archive size and availability of telomere-proximal *VSG*, possibly increasing the pool of telomere-proximal *VSG* that can interact readily with the expression site (switching) or with each other (recombination-mediated diversification).

Adaptations in nuclear structure and function. Trypanosome telomeric regions, including *VSG* genes, are located in heterochromatin at the nuclear periphery.[24] *VSG* transcription is mediated by RNA polymerase I (RNAPI), which normally transcribes only ribosomal RNA genes located in the nucleolus. The active *VSG* occupies the ES Body, an RNAPI-containing extranucleolar niche located within the nuclear interior rather than peripherally.[25] The minichromosomes behave differently from the core chromosomes, lying in the nuclear periphery, being duplicated earlier, and segregating via core microtubules rather than the spindle; the unusual behavior might be necessitated, at least partially, by overloading of the conventional machinery.[25]

Recombination prone sequences. Recombination and repair pathways serve fundamental functions in any organism, but they can be exploited to achieve specific phenotypes, such as in antigenic variation. One common mechanism is that an elevated rate of recombination causes the expressed gene to become replaced or altered. The elevated rate is specific to the contingency loci and is facilitated by the presence of recombination-prone sequences.[26,27] In the *VSG* system, the unstable, 70-bp repeats lying upstream of the expressed *VSG* are thought to precipitate a high level of recombination (see below). It is also apparent that the subtelomere compartment in the trypanosome genome differs from the chromosome cores in its interaction with recombination/repair mechanisms, becoming disproportionately disrupted in null mutants of key players MRE11 and BRCA2.[28,29] As to which recombination/repair pathways mediate *VSG* interactions, some interesting novelties exist in the trypanosome general pathways, but a formal link has yet to be made.[30]

Implicit information: pseudogenes and the combinatorial creation of novel intact genes. Some two-thirds of the array *VSGs* are pseudogenes, with premature stop codons, frameshifts, deletions, or lack of an appropriate start codon.[14] At first sight, this situation appears to be a colossal waste of resources, but in fact it is probably the opposite. By what is likely to be nonreciprocal recombination (gene conversion) of wild-type fragments of pseudogenes into the expressed *VSG*, antigenically distinct genes can be generated.[31] The pseudogenes

therefore encode implicit information.[32] Combinatorial use of silent information has potential for greatly expanding the extent of resulting, explicit information, as is demonstrated in the case of the bacterium *Anaplasma*, which can create apparently hundreds of antigenically distinct MSP2 variants from an archive of merely five unique pseudogenes, and no intact genes.[33] The trypanosome pseudogene pool is more than two orders of magnitude greater in size. That short conversion fragments can introduce novel function is apparent also from the human HLA and KIR cell surface proteins of the human immune system.[34–36]

How the antigenic variation phenotype is served by genome adaptations (Table 1)
Singular VSG expression. The ES Body provides expression to one ES at a time, and a telomere position effect, possibly along with other epigenetic effects, provides silencing of the others, ensuring controlled expression of the archive.

Table 1. Genome features contributing to the antigenic variation phenotype.[a]

Phenotype	Genome feature
Many VSGs	Large subtelomeres populated by *VSG* arrays
Singular VSG expression	ES Body: specialized nuclear niche; silencing through telomere position effects
Switching VSG	Telomere proximity of expression site enables interaction with *VSG* in other (sub)telomeres
Ordered expression—general	Locus types within subtelomeres have different activation probabilities
Ordered expression—specific	Flanks of each *VSG*? For mosaic *VSG* expression, coding sequence identity between donor and expressed genes
Antigenically diverse growth peaks	Blocks of distinct *VSG* with similar activation probability
Coordinated growth, differentiation, and antigenic variation	Number of *VSG* constituting a block. Higher number tends toward differentiation-based control; lower number favors antibody-mediated control
Prolonged infection	Mechanisms for ordered expression Mosaic *VSG* formation
Reinfection, superinfection	Diversity in *MVSG* set expressed in infective, metacyclic population? Mosaic *VSG* formation Contents and organization of blocks?
Host range	Large size of archive and its organization into blocks allow it to be used flexibly in different within-host environments, providing optimum transmission phenotype in each host species?
VSG hyperevolution	Enhanced mutation/reduced mutation repair in subtelomeres

[a]Speculation is denoted by question marks.

Switching arises possibly from inherent tendency for double-strand breaks. The 70-bp repeats upstream of the transcribed *VSG* each contain a tract of the physically unstable motif $(TAA)_n$.[37] Instability of this type theoretically can cause stalling of DNA replication, leading in effect to production of a double-strand break (DSB) in the newly synthesized duplex and, consequently, induction of repair.[38] A common route taken is creation of a gap and filling by copying from a homologous sequence. In this case, homology at another run of 70-bp repeats, upstream of a different *VSG*, would initiate the copying of that gene into the expression site. There is indeed some evidence that artificial creation of a DSB adjacent to the 70-bp repeats prompts switching by duplication, and that DSB occur naturally in the repeats in the expression site.[39] As with other contingency loci systems, where runs of short repeats are thought to cause gene inactivation or reactivation through indels arising by DNA polymerase slippage on the repeat template,[8] this mechanism is simple—an accident waiting to happen, at reasonably high frequency, due to inherent instability in the DNA sequence.

Ordered VSG expression. Antigenic variation runs in a pattern often termed *semipredictable*, with variants tending to appear in the same general order in distinct infections.[1] Accomplishing order for such an enormous archive seems inordinately difficult, but it is facilitated in part by the genome environment and structure. Each gene appears to have an inherent activation probability. Those with the highest probability will activate continuously, but early onset of antibodies against their product will render all subsequent activations futile. The process will repeat as the activation hierarchy is unrolled. In general, telomere-proximal genes have the highest activation probability, followed by intact genes in the arrays, and then pseudogenes.[1] Presumably, the BES telomere will interact more frequently with a donor at another telomere, via standard telomere associations, than with a donor in the subtelomere, which will require more routine homology searching. When mosaic gene formation comes into play, the interaction of donor with the BES is driven by degree of identity in the *VSG* coding sequence, making switching dependent on what went before.[14]

This gross level of order is refined, with blocks of variants appearing within each locus category.

Whereas the degree of coding identity will specify finer order for mosaic gene formation, we do not know how refinement is achieved for the other genes. A clue, however, lies in the *Borrelia* antigenic variation system, where fairly exquisite ordering is achieved via two sequences, one in either flank of the gene.[40] It is conceivable that the conserved sequences flanking *VSG* might have similar influence.

Diverse VSG expression. Each peak comprises a mixture of variants, the number of which is probably critical to the balance between antibody-mediated and density-dependent limitation of growth.[3] In addition, diversity within each of these blocks decreases the extent to which antibody responses might eliminate the infection. Diversity probably occurs by all variants in a block having very similar activation probabilities.

Prolonged infection, reinfection, and superinfection. The combination of order and enormous potential for combinatorial creation of novel, mosaic expressed *VSG* can lead, at least in theory, to infections lasting months or even years. It has been hypothesized that the combinatorial mechanism also can allow reinfection of hosts already immune to that strain.[11] Experimental analysis has shown that mosaic gene formation does enable reinfection of calves with the bacterium *Anaplasma*.[41] Comparative genomics has shown that *VSG* archives are strain specific,[42] a situation that enables strains to compete with each other in coinfections. In turn, competition will drive toward greater evolution of the archive.

Hyperevolution: selection of mechanisms that generate broad diversity. Second-order (indirect) selection[43] underpins evolution of antigenic variation. Every *VSG* is dispensable, and most, being very rarely expressed, have little exposure to direct selection via their encoded protein. Yet, diversity appears to be spread evenly over the archive, for example with no evident outgrouping of the high-probability activators. Typically for multigene families in subtelomeres, the archive evolves rapidly. It seems that second-order selective processes operate, in which mutational mechanisms generate diversity across the whole *VSG* archive and across the whole coding sequence within each gene, independently of the immediate phenotype;[14,42,44,45] it is these mechanisms that have been selected. Thus, an individual trypanosome with a novel trait that enhances population survival will be selected, but the dominant selective pressure acts at the level of the lineage, yielding mechanisms that directly confer benefit on the population rather than the individual. The view that evolution acts on only the immediate phenotype, rather than on long-term strategy is not compatible with what is observed for trypanosomes.

It is likely that the subtelomere provides an ideal environment for hypermutation, through naturally ectopic interactions and possibly through differential positioning within the nucleus. It is more likely that standard recombination-repair activities are used differentially on subtelomeres than that exclusive mutagenic mechanisms have evolved. *VSG* genes mutate in various ways. Ectopic gene conversion events duplicate *VSG* between subtelomeres, either as blocks of genes or individually.[14] Recently, comparative genomics of sequential isolates of one trypanosome strain have revealed a range of mutations, including base substitutions, short indels, and conversions within the coding sequence (L.P., T. Otto, M. Berriman, and J.D.B., unpublished).

Summary

To return to our question of how a parasite, with a smaller and less complex genome than that of its mammalian host, can compete in the face of antibody population diversity, the answer is flexibility. The trypanosome genome is adapted in somewhat extreme ways to serve the antigenic variation phenotype, with approximately 25% of the genome devoted to this phenomenon. The special features of subtelomeres, including hyperevolution and reversible gene-silencing mechanisms, have been exploited to great effect, resulting in an enormous archive of rapidly mutating silent *VSG* genes and telomere-associated mechanisms for their differential expression. Subtelomeres containing *VSG* arrays have expanded, and a large set of supernumerary chromosomes has emerged, apparently to provide even more telomeres. These structural adaptations have required changes in nuclear machinery. As with enhanced phenotypic variation systems in other pathogens, often referred to as *contingency gene systems*, the spontaneous phenotype switches that underpin antigenic variation appear to occur through inherent instability of a DNA sequence upstream of *VSG*. The silent *VSG* archive evolves at a rate several times faster than genes in chromosome cores, and it appears to do so without direct selection on expressed *VSG*. Instead, in a strong example of second-order selection, mechanisms have evolved that mutate all *VSG* apparently randomly. Much of the coding information is stored in the genome implicitly, in pseudogenes, requiring formation of mosaic genes for expression of novel antigens. The trypanosome *VSG* and mammalian antibody phenotypes both derive very extensive augmentation of information through the power of combinatorial reassembly of coding sequence. Besides these adaptations at the level of the genome, management of antigenic variation requires networking of *VSG* gene expression with other trypanosome phenotypes, including growth, differentiation, and, beyond the individual infection, unpredictable transmission into a very broad range of hosts, some of which will be partially immune.

Acknowledgment

This work was supported by the Wellcome Trust (Grant numbers 055558, 083224, and 086415). The Wellcome Trust Centre for Molecular Parasitology is supported by core funding from the Wellcome Trust (Grant number 085349).

Conflicts of interest

The authors declare no conflicts of interest.

References

1. Morrison, L.J., L. Marcello & R. McCulloch. 2009. Antigenic variation in the African trypanosome: molecular mechanisms and phenotypic complexity. *Cell Microbiol.* **11:** 1724–1734.
2. Kenter, A.L., S. Feldman, R. Wuerffel, *et al.* 2012. Three-dimensional architecture of the IgH locus facilitates class switch recombination. *Ann. N.Y. Acad. Sci.* **1267:** 86–94. This volume.
3. Gjini, E. *et al.* 2010. Critical interplay between parasite differentiation, host immunity, and antigenic variation in trypanosome infections. *Am. Nat.* **176:** 424–439.
4. Barry, J.D. & R. McCulloch. 2001. Antigenic variation in trypanosomes: enhanced phenotypic variation in a eukaryotic parasite. *Adv. Parasitol.* **49:** 1–70.
5. La Greca, F. & S. Magez. 2011. Vaccination against trypanosomiasis: can it be done or is the trypanosome truly the ultimate immune destroyer and escape artist? *Hum Vaccin.* **7:** 1225–1233.
6. Schwede, A. & M. Carrington. 2010. Bloodstream form Trypanosome plasma membrane proteins: antigenic variation and invariant antigens. *Parasitology* **137:** 2029–2039.

7. Barry, J.D. 1986. Antigenic variation during *Trypanosoma vivax* infections of different host species. *Parasitology* **92:** 51–65.

8. Bayliss, C.D. & M.E. Palmer. 2012. Evolution of simple sequence repeat mediated phase variation in bacterial genomes. *Ann. N.Y. Acad. Sci.* **1267:** 39–44. This volume.

9. MacGregor, P. *et al.* 2011. Transmission stages dominate trypanosome within-host dynamics during chronic infections. *Cell Host Microbe* **9:** 310–318.

10. Lythgoe, K.A. *et al.* 2007. Parasite-intrinsic factors can explain ordered progression of trypanosome antigenic variation. *Proc. Natl. Acad. Sci. USA* **104:** 8095–8100.

11. Barry, J.D. *et al.* 2005. What the genome sequence is revealing about trypanosome antigenic variation. *Bioch. Soc. Trans.* **33:** 986–989.

12. Ginger, M.L. *et al.* 2002. Ex vivo and in vitro identification of a consensus promoter for VSG genes expressed by metacyclic-stage trypanosomes in the tsetse fly. *Eukaryot Cell* **1:** 1000–1009.

13. Berriman, M. *et al.* 2005. The genome of the African trypanosome *Trypanosoma brucei*. *Science* **309:** 416–422.

14. Marcello, L. & J.D. Barry. 2007. Analysis of the VSG gene silent archive in *Trypanosoma brucei* reveals that mosaic gene expression is prominent in antigenic variation and is favored by archive substructure. *Genome Res.* **17:** 1344–1352.

15. Barry, J.D. *et al.* 2003. Why are parasite contingency genes often associated with telomeres? *Int. J. Parasitol.* **33:** 29–45.

16. Riethman, H., A. Ambrosini & S. Paul. 2005. Human subtelomere structure and variation. *Chromosome Res.* **13:** 505–515.

17. Callejas, S. *et al.* 2006. Hemizygous subtelomeres of an African trypanosome chromosome may account for over 75% of chromosome length. *Genome Res.* **16:** 1109–1118.

18. Therizols, P. *et al.* 2010. Chromosome arm length and nuclear constraints determine the dynamic relationship of yeast subtelomeres. *Proc. Natl. Acad. Sci. USA* **107:** 2025–2030.

19. Koszul, R., B. Dujon & G. Fischer. 2006. Stability of large segmental duplications in the yeast genome. *Genetics* **172:** 2211–2222.

20. Hertz-Fowler, C. *et al.* 2008. Telomeric expression sites are highly conserved in *Trypanosoma brucei*. *PLoS ONE* **3:** e3527.

21. Vink, C., G. Rudenko & H.S. Seifert. 2011. Microbial antigenic variation mediated by homologous DNA recombination. *FEMS Microbiol. Rev.* Dec 25. doi:10.1111/j.1574-6976.2011.00321.x. [Epub ahead of print].

22. Yang, X. *et al.* 2009. RAP1 is essential for silencing telomeric variant surface glycoprotein genes in Trypanosoma brucei. *Cell* **137:** 99–109.

23. Wickstead, B., K. Ersfeld & K. Gull. 2004. The small chromosomes of *Trypanosoma brucei* involved in antigenic variation are constructed around repetitive palindromes. *Genome Res.* **14:** 1014–1024.

24. DuBois, K.N. *et al.* 2012. NUP-1 Is a large coiled-coil nucleoskeletal protein in trypanosomes with lamin-like functions. *PLoS Biol.* **10:** e1001287.

25. Daniels, J.P., K. Gull & B. Wickstead. 2010. Cell biology of the trypanosome genome. *Microbiol. Mol. Biol. Rev.* **74:** 552–569.

26. Fondon, J.W., 3rd *et al.* 2008. Simple sequence repeats: genetic modulators of brain function and behavior. *Trends Neurosci.* **31:** 328–334.

27. Gemayel, R. *et al.* 2010. Variable tandem repeats accelerate evolution of coding and regulatory sequences. *Annu. Rev. Genet.* **44:** 445–477.

28. Robinson, N.P. *et al.* 2002. Inactivation of Mre11 does not affect VSG gene duplication mediated by homologous recombination in *Trypanosoma brucei*. *J. Biol. Chem.* **277:** 26185–26193.

29. Hartley, C.L. & R. McCulloch. 2008. Trypanosoma brucei BRCA2 acts in antigenic variation and has undergone a recent expansion in BRC repeat number that is important during homologous recombination. *Mol. Microbiol.* **68:** 1237–1251.

30. Horn, D. & R. McCulloch. 2010. Molecular mechanisms underlying the control of antigenic variation in African trypanosomes. *Curr. Opin. Microbiol.* **13:** 700–705.

31. Barbet, A.F. & S.M. Kamper. 1993. The importance of mosaic genes to trypanosome survival. *Parasitol. Today* **9:** 63–66.

32. Caporale, L.H. 2006. *The Implicit Genome*. Oxford University Press. Oxford.

33. Futse, J.E. *et al.* 2005. Structural basis for segmental gene conversion in generation of *Anaplasma marginale* outer membrane protein variants. *Mol. Microbiol.* **57:** 212–221.

34. Parham, P., E.J. Adams & K.L. Arnett. 1995. The origins of HLA-A,B,C polymorphism. *Immunol. Rev.* **143:** 141–180.

35. Ohta, T. 2010. Gene conversion and evolution of gene families: an overview. *Genes* **1:** 349–356.

36. Graef, T. *et al.* 2009. KIR2DS4 is a product of gene conversion with KIR3DL2 that introduced specificity for HLA-A*11 while diminishing avidity for HLA-C. *J. Exp. Med.* **206:** 2557–2572.

37. Ohshima, K. *et al.* 1996. TTA.TAA triplet repeats in plasmids form a non-H bonded structure. *J. Biol. Chem.* **271:** 16784–16791.

38. Labib, K. & B. Hodgson. 2007. Replication fork barriers: pausing for a break or stalling for time? *EMBO Rep.* **8:** 346–353.

39. Boothroyd, C.E. *et al.* 2009. A yeast-endonuclease-generated DNA break induces antigenic switching in Trypanosoma brucei. *Nature* **459:** 278–281.

40. Barbour, A.G. *et al.* 2006. Pathogen escape from host immunity by a genome program for antigenic variation. *Proc. Natl. Acad. Sci. USA* **103:** 18290–18295.

41. Futse, J.E. *et al.* 2008. Superinfection as a driver of genomic diversification in antigenically variant pathogens. *Proc. Natl. Acad. Sci. USA* **105:** 2123–2127.

42. Hutchinson, O.C. *et al.* 2007. Variant Surface Glycoprotein gene repertoires in Trypanosoma brucei have diverged to become strain-specific. *BMC Genomics* **8:** 234.

43. King, D.G. & Y. Kashi. 2007. Indirect selection for mutability. *Heredity* **99:** 123–124.

44. Jackson, A.P. *et al.* 2012. Antigenic diversity is generated by distinct evolutionary mechanisms in African trypanosome species. *Proc. Natl. Acad. Sci. USA* **109:** 3416–3421.

45. Oliveira, B.M., M. Watkins, P. Bandyopadhyay, *et al.* 2012. Adaptive radiation of venomous marine snail lineages and the accelerated evolution of venom peptide genes. *Ann. N.Y. Acad. Sci.* **1267:** 61–70. This volume.

Ann. N.Y. Acad. Sci. ISSN 0077-8923

ANNALS OF THE NEW YORK ACADEMY OF SCIENCES

Issue: *Effects of Genome Structure and Sequence on Variation and Evolution*

The tricky path to recombining X and Y chromosomes in meiosis

Liisa Kauppi,[1] Maria Jasin,[2] and Scott Keeney[1,3]

[1]Molecular Biology Program, [2]Developmental Biology Program, [3]Howard Hughes Medical Institute, Memorial Sloan-Kettering Cancer Center, New York, New York

Address for correspondence: Liisa Kauppi, Molecular Biology Program, Box 97, Memorial Sloan-Kettering Cancer Center, 1275 York Avenue, New York, NY 10065. KauppiL@mskcc.org

Sex chromosomes are the Achilles' heel of male meiosis in mammals. Mis-segregation of the X and Y chromosomes leads to sex chromosome aneuploidies, with clinical outcomes such as infertility and Klinefelter syndrome. Successful meiotic divisions require that all chromosomes find their homologous partner and achieve recombination and pairing. Sex chromosomes in males of many species have only a small region of homology (the pseudoautosomal region, PAR) that enables pairing. Until recently, little was known about the dynamics of recombination and pairing within mammalian X and Y PARs. Here, we review our recent findings on PAR behavior in mouse meiosis. We uncovered unexpected differences between autosomal chromosomes and the X–Y chromosome pair, namely that PAR recombination and pairing occurs later, and is under different genetic control. These findings imply that spermatocytes have evolved distinct strategies that ensure successful X–Y recombination and chromosome segregation.

Keywords: meiosis; recombination; sex chromosomes; chromosome pairing

In higher eukaryotes, meiotic cell divisions produce sperm and eggs. Haploid DNA content in germ cells is achieved by one round of DNA replication, followed by two successive rounds of cell division. To successfully segregate chromosomes to daughter cells in the first meiotic cell division, homologous chromosomes (homologs) must find each other and stably pair. In most organisms studied to date, including mammals, homolog recognition and pairing is achieved by DNA recombination interactions that are initiated by developmentally programmed DNA double-stranded breaks (DSBs).[1]

Evolutionary geneticists have long appreciated the role of meiotic recombination in reshuffling genetic material from one generation to the next (see, e.g., Ref. 2). Gene mapping efforts are possible thanks only to this feature of meiosis. What has received less attention in this field is the fact that meiotic recombination also has an essential function in chromosome pairing[3–5] and in establishing a physical connection (crossover) between homologs,[6,7]

which in turn is critical for chromosome segregation. To initiate recombination, mouse spermatocytes typically make 200–250 DSBs per nucleus.[8,9] As meiosis progresses, most of these DSBs are repaired without the exchange of flanking chromosome arms (noncrossover), while a subset matures into crossovers involving a reciprocal exchange between homologs[10] (see also Ref. 11). Recombination failure leads to inviable gametes.[3,5]

Once a DSB is made, the proper partner—the unbroken template for homologous recombination—must be located and engaged in a stable pairing interaction. Not all chromosomes, however, are on a level playing field with regard to pairing. Autosomal chromosomes are homologous along their entire length ($\sim$60–200 Mb, depending on the chromosome) and are hence able to pair from end to end. The X and Y chromosomes, on the other hand, are nonhomologous save for a short segment called the pseudoautosomal region (PAR).[12] X–Y chromosome homology search and pairing therefore

doi: 10.1111/j.1749-6632.2012.06593.x

can only be mediated by DSBs in the PAR, conceivably making X–Y segregation particularly difficult. Indeed, X–Y aneuploidy is a common paternally inherited chromosome disorder.[13,14] In most cells, however, X–Y chromosome segregation is completed successfully, raising the question as to how the challenging task of recombination between heteromorphic sex chromosomes is achieved. We applied sophisticated cytological methods to investigate DSB formation and pairing of the PARs in mice, and review here our recent findings on mechanisms that secure faithful X–Y recombination.[15]

Unusual higher-order chromatin structure in the PAR

Ensuring that at least one PAR DSB is formed presents the first potential hurdle on the path to successful X–Y recombination: in mice, one meiotic DSB forms every 10 Mb on average,[16] while the PAR spans just 0.7 Mb.[17] How do spermatocytes promote a DSB frequency 10- to 20-fold higher than average in this tiny region? Insight into this question came from consideration of higher-order chromosome structure. Meiotic DSB formation takes place in the context of highly organized chromatin;[18] replicated sister chromatids emanate from chromosome axes as chromatin loops (Fig. 1). Blat *et al.*[19] proposed a model in which chromatin loops are accessible to the DSB machinery, with each loop therefore presenting an opportunity for DSB formation. Thus, DNA

packaged into many small loops may be more prone to DSB formation than if the same kilobase-amount of DNA was packaged into fewer and larger loops. We therefore speculated that one way to render the PAR more conducive to DSB formation could be packaging it into larger numbers of smaller loops.

To gain insight into the higher-order structure of the PAR, we measured how far a fluorescent *in situ* hybridization (FISH) probe signal extended from chromosome axes, compared to probes on small autosomes, in nuclei isolated from spermatocytes in early meiosis. Autosomal probes extended on average > five-fold further than the PAR probe (Fig. 1A).[15] This observation, combined with the fact that PAR chromosome axes are disproportionately long considering their DNA content (Fig. 1B), indicates that PAR chromatin loops are several-fold smaller than loops in autosomal genomic regions (Fig. 1C), potentially providing the meiotic DSB machinery with increased opportunity for cutting. Hence, distinct chromatin packaging may be the first level of control that helps promote high-frequency DSB formation in the PAR.

Most DSBs in PARs form later than on autosomes

We asked whether PAR DSBs form with the same timing as "bulk" (nucleus-wide) DSBs. To answer this question, we examined meiotic chromosome preparations of nuclei at various stages of

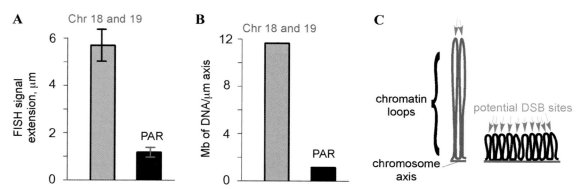

Figure 1. Higher-order structure of the PAR compared to autosomes. (A) Measurements of the maximal distance of FISH signals from chromosome axes (average for subtelomeric chromosome 18 and 19 probes, gray bar; PAR probe, black bar) in cells at pachynema (mean ± SD). (B) DNA content (Mb) per μm of chromosome axis in the subtelomeric regions of chromosomes 18 and 19, and the PAR. (C) Model of how organization of DNA into chromatin loops can influence DSB frequency (after Ref. 19). Only one homolog (each consisting of a pair of sister chromatids) is shown, with chromatin loops tethered to chromosome axes (red lines). Each chromatin loop is envisioned to provide an opportunity for meiotic DSB formation (orange arrows). DNA organized into more and smaller loops per Mb (right) presents the DSB machinery with more opportunities for break formation than if the same length of DNA is organized into fewer and larger loops (left).

meiotic progression. Stages are defined by the extent of chromosome axis development, assessed by the synaptonemal complex protein SYCP3, used as a cytological marker.[15] We first performed immunofluorescence (IF) against SYCP3 (to visualize meiotic chromosome axes) and RAD51 (which

marks sites of DSB repair), to quantify nucleus-wide DSBs (Fig. 2A, i). IF was followed by PAR FISH (Fig. 2A, ii) to examine whether PARs displayed a RAD51 focus (Fig. 2A, iii). In control mice (blue circles in Fig. 2B), nucleus-wide RAD51 focus numbers peaked in early/mid-zygonema. Surprisingly, in the

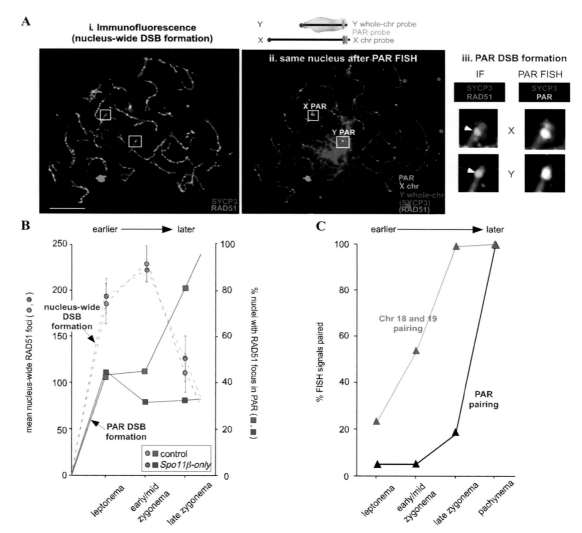

Figure 2. Many PAR DSBs form later than nucleus-wide DSBs, and are under different genetic control. (A) Assay for nucleus-wide and PAR DSB formation (i–ii). Example of IF and FISH staining of a spermatocyte in late zygonema. Nuclei were first stained with antibodies against RAD51 and SYCP3 and photographed (i); then FISH was performed using PAR, X and Y probes (depicted above the FISH image), and the same nuclei were photographed again (ii).[14] Under the FISH conditions used in these experiments, some IF signal remains visible. The presence or absence of a RAD51 focus in the X and Y PAR was scored by comparing the IF image to an IF+FISH image overlay (iii). In this example, both the X and Y PAR (inside white squares in i–ii) display a RAD51 focus (white arrows in iii). Scale bar, 10 μm. (B) Mean nucleus-wide RAD51 focus numbers (bars ± 95% CI of the mean; left Y axis) and percentage of nuclei with a PAR RAD51 focus on the X and/or Y PAR (right Y axis) as prophase I progresses from leptonema to late zygonema, in mice of the indicated genotypes (see text). (C) Percentage of nuclei with paired FISH signals for autosomal and PAR probes as prophase I progresses from leptonema to pachynema.

same mice, PAR RAD51 foci were not observed until late zygonema in most nuclei (blue squares in Fig. 2B). This analysis indicated that many cells do not experience a PAR DSB until late zygonema, whereas nucleus-wide DSBs reach a maximum earlier, in early/mid-zygonema (Fig. 2B, compare blue solid line with blue dashed line). For data and in-depth discussion of RAD51 foci observed on the X PAR versus the Y PAR, see Ref. 15.

PAR DSB formation is under distinct genetic control

The temporally distinct (delayed) DSB formation of the PARs raised the possibility of differential genetic control. The protein responsible for meiotic DSB formation, SPO11,[20] has two major isoforms in mammals, SPO11α and SPO11β.[21–23] Interestingly, the two isoforms appear with differential timing: SPO11β predominates in early meiotic cells when most DSBs form, whereas SPO11α is more abundant later.[15,23] It therefore seemed possible that the late isoform, SPO11α, could play a role in coordinating later-forming DSBs in general, or PAR DSBs in particular. To elucidate the genetic requirements for nucleus-wide and PAR DSB formation, we analyzed RAD51 foci in a transgenic mouse model carrying only the Spo11β isoform (*Spo11β*-only mice).[15]

Nucleus-wide ("global") RAD51 focus numbers in *Spo11β*-only mice were indistinguishable from those in control mice (Fig. 2B, compare red dashed line with blue dashed line), suggesting that SPO11β by itself is sufficient for normal overall DSB levels. In contrast, the percentage of cells with PAR RAD51 foci was significantly lower at late zygonema, the stage when most PARs in control mice display a RAD51 focus (Fig. 2B, compare red solid line with blue solid line). Late PAR DSBs therefore are genetically separable from both nucleus-wide DSBs and early PAR DSBs. The absence of SPO11α apparently affects only late-forming PAR DSBs. The cellular consequences are discussed in the next section.

X and Y chromosomes pair later than autosomes

Given our observation that PAR DSBs form later than autosomal DSBs, one would predict that PAR (that is, X and Y chromosome) pairing would occur later than autosomal pairing in wild-type mice. We assessed the pairing status of PARs and autosomes by asking whether cells at various stages of

meiosis displayed two (i.e., unpaired) or one (i.e., paired) FISH probe signals. We discovered that the X and Y PARs indeed pair late in comparison to the autosomes we examined (Fig. 2C). Contemporaneously, and independent of DSB formation, a heterochromatic domain forms called the *sex body*, which brings the X and Y chromosomes closer together.[24] This physical proximity of the X and Y at late zygonema likely is an additional (but not sufficient) factor that facilitates PAR pairing.

A further prediction is that since PAR DSBs fail to form in many *Spo11β*-only spermatocytes, PAR pairing should also often fail. This is exactly what we found: 69% of PARs were unpaired in pachynema in *Spo11β*-only mice, compared to 5% in control mice; in contrast, no change in autosomal pairing was found.[15] As *Spo11β*-only spermatocytes progress to metaphase I, premature X–Y separation was observed in a similarly high percentage of cells. Massive apoptosis was seen in the testes at this stage and mice are largely infertile.[5] The simplest explanation for these meiotic defects is that in the majority of cells, PAR DSB formation and subsequently crossover formation fails, leading to unconnected X and Y chromosomes at metaphase whose presence triggers the spindle checkpoint and an apoptotic response.

Interestingly, we found that *Spo11β*-only female mice are fertile, and all chromosomes (including the fully homologous X chromosomes) pair normally.[15] Therefore in female meiosis, although SPO11α is expressed,[15,22] it appears to be entirely dispensable. If SPO11α's main function indeed is to ensure efficient PAR DSB formation, the *Spo11β*-only female fertility phenotype also indicates that PAR DSBs are dispensable for X–X pairing. This is perhaps not surprising, since recombination in females can take place along the nearly 170 Mb length of the X chromosome, instead of being restricted to the < 1-Mb PAR.

Conclusions and future directions

Our experiments demonstrated that PAR DSB formation, recombination, and pairing are temporally and genetically distinct from that of autosomal chromosomes. A model consolidating these findings is shown in Figure 3. Autosomes undergo DSB formation earlier, and pair earlier than PARs. Not until autosomal pairing is nearly complete, do most cells form a DSB in the PAR. Sex body formation likely

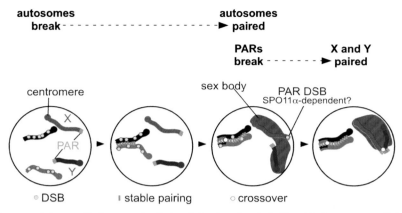

Figure 3. Model summarizing our findings on the distinct behavior of meiotic X and Y chromosomes.[15] A pair of autosomal homologous chromosomes is shown as light and dark gray lines, the X and Y as red and blue lines, respectively. Most DSBs on autosomes form before PAR DSBs; as a consequence, autosomal chromosomes pair earlier than the X and Y. Only DSBs that can facilitate homolog pairing are depicted (yellow circles); not shown are the numerous DSBs that form on the non-PAR portion of the X chromosome (see Ref. 15) but cannot mediate pairing. In many cells, PARs undergo DSB formation later, around the same time as the sex body (hatched area) begins to form. This chromatin domain brings the X and Y closer together and likely facilitates PAR pairing.

aids in PAR pairing by gathering the X and Y chromosomes into the same nuclear territory. In transgenic mice that lack SPO11α (the late-appearing isoform), late PAR DSBs are not formed, and PAR pairing frequently fails, leading to X–Y missegregation and male infertility. In a subset of cells, PAR DSBs form earlier, around the same time as nucleus-wide DSBs; these cells do not need SPO11α for pairing or crossing over. The simplest interpretation is that the main and perhaps only role of SPO11α is to promote late PAR DSB formation, either directly or indirectly. A *Spo11α*-only mouse model, by itself and crossed with *Spo11β*-only mice, should shed light on this matter.

Heteromorphic sex chromosomes and PARs are a widespread genomic feature in the animal kingdom.[12] This raises the question whether the distinct properties of PARs that we uncovered in mouse meiosis are a frequently used solution to the sex chromosome recombination problem. Circumstantial evidence suggests that this may be the case, at least for mammals: humans have the same SPO11 isoforms as mice,[21,23] and human X and Y chromosomes have been reported to pair later than autosomes.[25] With the immuno-FISH methods established in mice, detailed cytological studies into human X–Y chromosome dynamics are now feasible for the first time.

Accurate DNA recombination, with DSBs as initiating DNA lesions, is critical for genome stability and for fertility. DSBs represent potentially toxic DNA damage that can lead to genome rearrangements if not faithfully repaired. Yet, in each meiotic cell, > 200 self-inflicted DSBs on average are made and subsequently processed in a remarkably error-free manner. Our data highlight multiple layers of exquisite control that underlie this fidelity. First, higher-order chromatin structure can have an impact on which genomic regions are accessible to the recombination machinery. Second, the position of chromosomes in three-dimensional space within the nucleus influences DNA transactions. Third, the expression of proteins that interact with germline DNA is under stringent temporal regulation and restricts DSB formation to a limited window of opportunity. Future studies will further elucidate factors that affect genome plasticity. Nowhere is the repair of DNA damage more important than in the germline, where it has the potential to give rise to heritable changes and thereby to provide fuel for evolution.

Conflicts of interest

The authors declare no conflicts of interest.

References

1. Zickler, D. & N. Kleckner. 1999. Meiotic chromosomes: integrating structure and function. *Annu. Rev. Genet.* **33:** 603–754.

2. Barton, N.H. & B. Charlesworth. 1998. Why sex and recombination? *Science* **281:** 1986–1990.

3. Baudat, F. *et al.* 2000. Chromosome synapsis defects and sexually dimorphic meiotic progression in mice lacking Spo11. *Mol. Cell* **6:** 989–998.

4. Peoples, T.L. *et al.* 2002. Close, stable homolog juxtaposition during meiosis in budding yeast is dependent on meiotic recombination, occurs independently of synapsis, and is distinct from DSB-independent pairing contacts. *Genes Dev.* **16:** 1682–1695.

5. Romanienko, P.J. & R.D. Camerini-Otero. 2000. The mouse Spo11 gene is required for meiotic chromosome synapsis. *Mol. Cell* **6:** 975–987.

6. Eaker, S. *et al.* 2002. Meiotic prophase abnormalities and metaphase cell death in MLH1-deficient mouse spermatocytes: insights into regulation of spermatogenic progress. *Dev. Biol.* **249:** 85–95.

7. Lipkin, S.M. *et al.* 2002. Meiotic arrest and aneuploidy in MLH3-deficient mice. *Nat. Genet.* **31:** 385–390.

8. Cherry, S.M. *et al.* 2007. The Mre11 complex influences DNA repair, synapsis, and crossing over in murine meiosis. *Curr. Biol.* **17:** 373–378.

9. Roig, I. *et al.* 2010. Mouse TRIP13/PCH2 is required for recombination and normal higher-order chromosome structure during meiosis. *PLoS Genet.* **6:** e1001062.

10. Cole, F., S. Keeney & M. Jasin. 2010. Comprehensive, fine-scale dissection of homologous recombination outcomes at a hot spot in mouse meiosis. *Mol Cell* **39:** 700–710.

11. Cole, F., S. Keeney & M. Jasin. 2012. Preaching about the converted: how meiotic gene conversion influences genomic diversity. *Ann. N.Y. Acad. Sci.* **1267:** 95–102. This volume.

12. Raudsepp, T. *et al.* 2012. The pseudoautosomal region and sex chromosome aneuploidies in domestic species. *Sex. Dev.* **6:** 72–83.

13. Shi, Q. *et al.* 2001. Single sperm typing demonstrates that reduced recombination is associated with the production of aneuploid 24,XY human sperm. *Am. J. Med. Genet.* **99:** 34–38.

14. Shi, Q. & R.H. Martin. 2001. Aneuploidy in human spermatozoa: FISH analysis in men with constitutional chromosomal abnormalities, and in infertile men. *Reproduction* **121:** 655–666.

15. Kauppi, L. *et al.* 2011. Distinct properties of the XY pseudoautosomal region crucial for male meiosis. *Science* **331:** 916–920.

16. Barchi, M. *et al.* 2008. ATM promotes the obligate XY crossover and both crossover control and chromosome axis integrity on autosomes. *PLoS Genet.* **4:** e1000076.

17. Perry, J. *et al.* 2001. A short pseudoautosomal region in laboratory mice. *Genome Res.* **11:** 1826–1832.

18. Panizza, S. *et al.* 2011. Spo11-accessory proteins link double-strand break sites to the chromosome axis in early meiotic recombination. *Cell* **146:** 372–383.

19. Blat, Y. *et al.* 2002. Physical and functional interactions among basic chromosome organizational features govern early steps of meiotic chiasma formation. *Cell* **111:** 791–802.

20. Keeney, S., C.N. Giroux & N. Kleckner. 1997. Meiosis-specific DNA double-strand breaks are catalyzed by Spo11, a member of a widely conserved protein family. *Cell* **88:** 375–384.

21. Keeney, S. *et al.* 1999. A mouse homolog of the Saccharomyces cerevisiae meiotic recombination DNA transesterase Spo11p. *Genomics* **61:** 170–182.

22. Bellani, M.A. *et al.* 2010. The expression profile of the major mouse SPO11 isoforms indicates that SPO11beta introduces double strand breaks and suggests that SPO11alpha has an additional role in prophase in both spermatocytes and oocytes. *Mol. Cell Biol.* **30:** 4391–4403.

23. Romanienko, P.J. & R.D. Camerini-Otero. 1999. Cloning, characterization, and localization of mouse and human SPO11. *Genomics* **61:** 156–169.

24. Burgoyne, P.S., S.K. Mahadevaiah & J.M. Turner. 2009. The consequences of asynapsis for mammalian meiosis. *Nat. Rev. Genet.* **10:** 207–216.

25. Rasmussen, S.W. & P.B. Holm. 1978. Human meiosis II. Chromosome pairing and recombination nodules in human spermatocytes. *Carlsberg Res. Comm.* **43:** 275–327.

Ann. N.Y. Acad. Sci. ISSN 0077-8923

ANNALS OF THE NEW YORK ACADEMY OF SCIENCES

Issue: *Effects of Genome Structure and Sequence on Variation and Evolution*

Sites of genetic instability in mitosis and cancer

Anne M. Casper, Danielle M. Rosen, and Kaveri D. Rajula

Department of Biology, Eastern Michigan University, Ypsilanti, Michigan

Address for correspondence: Anne M. Casper, Department of Biology, 391 Mark Jefferson Science Complex, Eastern Michigan University, Ypsilanti, MI 48197. anne.casper@emich.edu

Certain chromosomal regions called common fragile sites are prone to difficulty during replication. Many tumors have been shown to contain alterations at fragile sites. Several models have been proposed to explain why these sites are unstable. Here we describe work to investigate models of fragile site instability using a yeast artificial chromosome carrying human DNA from a common fragile site region. In addition, we describe a yeast system to investigate whether repair of breaks at a naturally occurring fragile site in yeast, FS2, involves mitotic recombination between homologous chromosomes, leading to loss of heterozygosity (LOH). Our initial evidence is that repair of yeast fragile site breaks does lead to LOH, suggesting that human fragile site breaks may similarly contribute to LOH in cancer. This work is focused on gaining understanding that may enable us to predict and prevent the situations and environments that promote genetic changes that contribute to tumor progression.

Keywords: fragile site; FRA3B; flexibility peak; mitotic crossover; loss of heterozygosity; cancer

Introduction

Genetic changes that alter the expression of genes that regulate cell growth or genes that maintain genomic integrity drive tumorogenesis. Although each tumor has a unique set of changes to cellular DNA, these alterations are not wholly random. Loss of heterozygosity (LOH) at particular tumor suppressor genes, such as p53, and amplification of certain oncogenes, such as MET, are frequently observed. The mechanisms that underlie these recurrent changes have long been a subject of focused research. One group of genomic loci that frequently are found to be unstable in cancer cells (i.e., associated with deletions, amplifications, and translocations), and that can drive tumorigenesis, comprises common fragile sites (CFSs). These sites are chromosomal loci that form gaps or breaks on metaphase chromosomes under conditions that partially inhibit replication.[1] In particular, CFSs are highly sensitive to inhibition of DNA polymerases (for more discussion of other particular genomic zones that are prone to mutation under stressful environments, the interested reader is directed to the review by Moore *et al.*[2])

In a recent meta-analysis of databases of tumor-associated genetic changes, Burrow *et al.*[3] found that more than half of the reported cancer translocations have at least one breakpoint in a chromosome band containing a CFS. In addition, there are many reports of deletions within tumor suppressor genes that harbor CFSs,[4,5] and of CFS mapping to the borders of oncogenic amplicons that appear to result from breakage–fusion–bridge cycles.[6–8] Mitotic sister chromatid exchange is frequently observed at CFSs,[9] which suggests that breaks at CFSs could potentially also drive LOH in cancer cells if the homologous chromosome is chosen for repair by recombination.[10,11] A variety of agents when combined with replication stress, including caffeine, cigarette smoke, pesticides, chemotherapeutic drugs, and hypoxic conditions, also have been reported to increase the frequency of gaps and breaks at CFSs, which may further drive cancer initiation or progression.[12]

Analyses and comparisons of CFSs to determine the reason for their instability have demonstrated that CFSs are slow to finish the process of replication.[13–16] Therefore, these loci apparently are difficult to replicate and are particularly sensitive to replication delay.

doi: 10.1111/j.1749-6632.2012.06592.x

Hypotheses proposed to explain why CFSs are unstable

Inhibition of replication causes uncoupling between helicase and polymerase proteins, resulting in the excessive accumulation of single-stranded DNA at the replication fork. This model proposes that DNA sequences at CFS are particularly prone to forming secondary structures in this *single-stranded DNA (ssDNA)*.[9] These secondary structures further stall replication, and may lead to breaks either directly, by cleavage of the structure, or indirectly as a result of broken anaphase bridges formed at these regions because replication does not complete prior to cell division.[17]

Fragile sites are located at the boundaries between early- and late-replicating zones of the DNA. Replication forks from earlier-replicating zones may pause in CFS regions.[18,19] When replication is delayed, these paused forks may be prone to collapse into a DNA break, or, the nearby late-replicating region may not complete replication prior to chromosome condensation, leading to a break at this site.

There is a relative lack of origin initiation events within CFS regions. Conditions that slow polymerase progress result in the cell dividing before replication of the CFS region is complete, leading to DNA breaks.[20]

Many CFS are located within genes that are very large and take a long time to transcribe. Collisions between the RNA transcription machinery and DNA polymerase lead to breaks.[21]

These four models partially overlap, and are not necessarily mutually exclusive. However, they make different predictions about the location of breaks within CFS regions (the CFSs that have been molecularly characterized are large, from 200 kb to a Mb or more, with breaks throughout[9]).

In particular, the first and second models suggest there may be certain sequences within CFS regions that are hotspots for breaks, such as sequences with high potential to form secondary structures and replication pause sites. For example, several CFSs have been reported to contain a greater density of *flexibility peaks* relative to non-CFS regions.[22,23] Flexibility peak is a term used to describe an AT-rich region of DNA that is characterized by the potential for high-twist angle between bases.[22,23] In the third

and fourth models above, no particular sequence within a CFS would be expected to be a hotspot, but the third model may predict that breaks would be more likely to occur in a region farthest from the site of replication initiation.

In the work described here, to investigate the mechanism of CFS-induced breaks, we ask whether the flexibility peaks that have been identified within human CFS FRA3B are hotspots of instability. Second, to explore the consequences of CFS breaks, we investigate whether repair of fragile site breaks drives LOH events due to mitotic homologous recombination.

To gather detailed data on exact break locations within CFSs, we used a yeast artificial chromosome (YAC) containing the human common fragile site FRA3B. This YAC does not contain any sequences required for yeast survival, and thus there is no selective pressure to retain particular regions of it. We modified the yeast carrying this YAC so that repair of breaks by telomere capping close to the break site is favored. Data described below suggest that break sites are not randomly distributed, but rather are clustered at the centromere-distal end of the FRA3B sequence insert.

To investigate mitotic homologous recombination, we take advantage of a naturally occurring yeast fragile site known as FS2 (fragile site 2). Similar to human CFS, recurrent breaks occur at FS2 under stressful conditions where replication is impaired.[24] Our results described below suggest that inhibition of yeast DNA polymerase stimulates mitotic recombination between homologus chromatids with reciprocal crossovers at FS2, resulting in LOH.

Mapping break locations in a YAC containing human FRA3B sequence

We chose to examine FRA3B, one of the CFS most frequently broken in human lymphocytes, which is located within *FHIT*, a tumor suppressor gene on human chromosome 3. *FHIT* is large, encompassing more than 1.5 Mb, and FRA3B is a ~200 kb region within this gene from approximately intron 3 through intron 5. CEPH cloning library YAC 850a6 contains 1.3 Mb of human sequence, including FHIT exon 1 through part of intron 5, encompassing the entire FRA3B region (Fig. 1).[25–28]

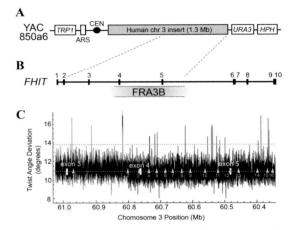

Figure 1. Structure and characteristics of YAC 850a6 and its human DNA insert. (A) YAC 850a6 carries a 1.3 Mb insert of DNA from human chromosome 3. The YAC also has a *TRP1* marker gene and a yeast origin of replication (ARS) on the left arm, and *URA3* and *HPH* markers on the right arm centromere-distal to the human DNA insert. (B) The human *FHIT* gene has ten exons, which are numbered and represented by short vertical bars. The FRA3B common fragile site is located within *FHIT*. The boundaries of this fragile site are not well defined, but there is general agreement that it is ~200 kb in size and spans from *FHIT* intron 4 through part of intron 5. Red dotted lines indicate the portion of the *FHIT*/FRA3B region that are carried on YAC 850a6. (C) DNA flexibility in the *FHIT* gene from exon 3 through the portion of intron 5 carried on the YAC was analyzed using a 100 bp sliding window. Regions with a twist-angle deviation over 13.7° (top dotted line) are considered flexibility peaks because they are more than 4.5 SD from the average flexibility.[22] The locations of primer sets 1–16 used in the analysis of broken YACs are shown as yellow arrows.

We chose to map break locations in YAC 850a6 following replication stress in yeast as this system has several features designed to simplify interpretation. First, since the fragile site FRA3B is carried on the extra, artificial chromosome rather than the yeast's own chromosome, there is no selection against loss of part of this chromosome to preserve essential yeast genes. In addition, because the chromosome is carried as a single copy in a haploid yeast cell, our analysis is not complicated by the presence of a homologous copy of the sequence. Also, because the genomic background of the yeast cell is easily modified, we can make modifications to prevent repair processes that obscure the YAC break site. We made two modifications to the yeast genome to favor repair of breaks by telomere capping close to the break site. First, we inserted the *pif1-m2* mutation, which increases the frequency of telomere-

capping.[29] Second, we deleted *EXO1*. Exo1p normally resects the 5′ end at double-strand breaks, therefore deletion of *EXO1* results in shorter resection tracts; these shorter tracts are expected to be preferable substrates for the telomere-capping mechanism.[30–32]

Human fragile sites break when cells are exposed to aphidicolin, a drug that inhibits DNA polymerases.[1] Although yeast are insensitive to aphidicolin, we can induce the stress of decreased DNA polymerase activity by employing a construct designed by Lemoine *et al.*,[24] in which polymerase alpha expression is controlled by the level of galactose in the media. In this construct, the GAL1/10 promoter drives expression of the *POL1* gene (encoding the catalytic subunit of polymerase alpha). When grown on medium containing low levels of galactose, cells with this *GAL-POL1* construct have low levels of polymerase alpha, and breaks are stimulated at a naturally occurring fragile site in yeast, FS2 (fragile site 2).[24] Thus, we used this system to induce breaks in the YAC carrying human FRA3B fragile site sequence by growing yeast cells on medium with low levels of galactose.

Genetic markers at the centromere-distal ends of each arm of the YAC facilitate identification of broken YACs by a phenotypic change (Fig. 1). The left arm is short and carries the *TRP1* marker gene and a yeast origin of replication (ARS), and the right arm carries the human DNA insert and the *URA3* marker gene. We also inserted another marker gene for hygromycin-resistance (*HPH*) distal to *URA3*. If a break occurs within the human DNA insert of the YAC, the cell retains *TRP1* but loses both the *URA3* and *HPH* genes. Loss of *URA3* results in resistance to 5-flourorotic acid (5-FOA), so cells with a broken YAC become auxotropic for tryptophan and uracil, resistant to 5-FOA, and sensitive to hygromycin. Genomic DNA from colonies with the phenotype indicating a broken YAC is evaluated by PCR using 16 primer sets spaced each 25–50 kb across the FRA3B insert (Fig. 1), with the break assumed to occur between the last primer set that amplified a product and the first set with no product.

In our initial data, in 23 independently isolated colonies, we identified breaks in the YAC only in the most centromere-distal region of the FRA3B insert, within *FHIT* intron 5:ten with a break between primer sets 13 and 14; three broken between sets 14 and 15; seven broken between sets 15 and 16; and three broken distal to primer set 16 (Fig. 1). Although FRA3B has been reported to exhibit breaks and gaps throughout a 200 kb region,[25,26] the clustered location of breaks in *FHIT* intron 5 that we observed in the YAC insert suggests that the tendency for break formation is not evenly distributed throughout the region, or that the process by which these breaks are repaired favors telomere capping in only this subregion of the sequence. We note that there are several peaks of high flexibility within the FRA3B DNA carried on the YAC. Mishmar *et al.*[22] created the FlexStab program for analysis of flexibility; this program evaluates local variation in DNA twist angle for dinucleotide pairs in sliding 100 bp windows, with values summed for the window and averaged for window length. Windows that are more than 4.5 SDs from the average flexibility are considered flexibility peaks. Mishmar[22] first reported that the human FRA7H region has a higher density of flexibility peaks than chromosomal bands lacking fragile sites, and later publications identified a similar pattern in other fragile sites,[9] leading to a shared hypothesis in the field that these flexibility peaks may instigate instability in fragile sites. However, the breakpoints we mapped are not clustered in regions with such peaks. Thus, our results to date suggest that flexibility peaks are not favored break sites, or that breaks do not occur at the site where polymerase is stalled. Because there are two flexibility peaks between primer sets 14 and 15, and two peaks between sets 15 and 16, we plan to design additional primers to further narrow the break locations between these primer sets to determine whether breaks in these regions occur at flexibility peaks, and/or at other DNA sequences with high secondary-structure–forming potential. It also is possible that flexibility peaks between primer sets 14 and 16 do stimulate breaks, but that the breakpoints mapped between primer sets 13 and 14 result from more extensive processing of the broken end by an exonuclease prior to telomere capping. Although we tried to avoid that by using cells mutant for *exo1*, further work is needed to examine whether other exonucleases, such as Sgs1p, might compensate in its absence.[30–32] Finally, it is important to note that the breakpoint locations we have mapped are all at the end of the human DNA insert that is farthest from the yeast origin on the YAC. This result may support the third model for fragile site instability (lack of origin activation). If there is

a lack of origin activation within the FRA3B region carried on the YAC, then the centromere-distal end of the insert is more likely to have difficulty completing DNA replication prior to cell division than the proximal end of the insert. However, this result may also support the first model for fragile site instability (secondary structure formation in extended ssDNA) if further narrowing of the break locations reveals that breaks have occurred at a specific sequence that has a strong potential for intrastrand secondary structure formation.

Investigating whether fragile site instability stimulates mitotic recombination causing LOH

LOH in genomic regions with tumor suppressor genes is an important driver of tumor initiation and progression.[33] Both deletion and mitotic recombination can result in LOH. Mitotic recombination has been understudied as a cause of LOH because it has been technically challenging to recover both cells following a recombination event and because in mitosis, recombination events between homologous chromosomes are rare relative to recombination between sister chromatids. To overcome these technical challenges, we chose to use a system recently developed in yeast that allows recovery of both cells following mitotic recombination between homologous chromosomes[10,11] (for highlights of a recently developed system to analyze all cell products following meiotic recombination events between homologs in a mouse model system, the interested reader is directed to the review by Cole *et al.*[34]).

The yeast model system developed for analysis of mitotic recombination between homologous chromosomes affords us the opportunity to study whether recombination events resulting in LOH are stimulated by instability at fragile sites. This system uses diploid yeast that has ∼ 0.5% sequence divergence between homologous chromosomes and which are homozygous for the *ade2–1* mutation.[11] This mutation results in cells that are Ade⁻ and red in color, because in the adenine biosynthesis pathway, the substrate bound by the Ade2p enzyme is a red molecule. The particular mutation in *ade2–1* is an ochre stop codon, which can be suppressed by the SUP4-o tRNA. By placing the *SUP4-o* gene on only one homolog of a chromosome of interest, the phenotype of the cell is then white and Ade⁺ (Fig. 2). When these cells are plated, mitotic

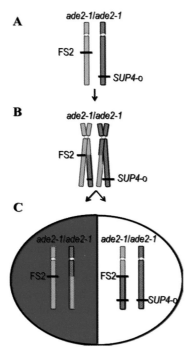

Figure 2. System for analysis of mitotic recombination between homologous chromosomes. (A) The starting yeast strain is diploid and is homozygous for the *ade2–1* mutation, which causes a block in the adenine biosynthesis pathway and accumulation of a red-pigmented intermediate molecule. Only yeast chromosome III from this diploid is shown. One copy of chromosome III contains the naturally occurring yeast fragile site FS2. The homologous copy of chromosome III does not contain this fragile site but does have the *SUP4*-o tRNA gene inserted as shown. This tRNA suppresses the *ade2–1* mutation, thus the starting strain is Ade+ and white in color. (B) A reciprocal crossover at the FS2 locus during mitotic cell division is shown. (C) In half of all reciprocal crossover events, chromosome segregation will result in the pattern shown, causing a red/white sectored colony in which chromosome III in each half of the sector is homozygous for all SNPs centromere-distal to the crossover. The left cell is red because the *ade2–1* mutation is not suppressed (no *SUP4*-o gene is present). As a result of the crossover, both copies of the *SUP4*-o gene are in the right cell, making it white. Other mitotic recombination events that result in red/white sectoring, not shown but discussed in the text, include gene conversion and break-induced replication.

recombination events that cause LOH at the *SUP4*-o locus will result in red/white sectored colonies (Fig. 2).[11] We have placed the *SUP4*-o gene on the distal end of the right arm of one homolog of chromosome III. This chromosome contains the naturally occurring yeast fragile site, FS2, which mimics the instability at CFS in human cells because breaks

at the site are stimulated by partial inhibition of replication.[24]

To study whether breaks at FS2 stimulate mitotic recombination events resulting in LOH, we inserted the *GAL-POL1* construct into yeast with the red/white detection system. As described earlier, when cells are grown in medium with low levels of galactose, this construct causes replication stress by lowering the level of polymerase alpha.[24] Cells are grown for six hours in low galactose media, and then plated for single colonies on medium with high galactose. Colonies that appear as red/white sectors are selected for further analysis. A single cell is purified from each half of the sector, and expanded for genomic DNA harvest. Since the starting diploid strain has ∼ 0.5% sequence divergence, single nucleotide polymorphisms that differ between the two homologous copies of chromosome III are evaluated to identify the type of event resulting in LOH, and the location on the chromosome where that event was initiated. For example, as illustrated in Figure 2, a reciprocal crossover (RCO) will result in homozygosity of all SNPs distal to the crossover, in both the red and white cells. In contrast, a gene conversion event that is not associated with a crossover will result in one cell remaining heterozygous at all SNPs, while the other will become homozygous for SNPs in the region of gene conversion. A break-induced replication event will result in one cell remaining heterozygous at all SNPs, while the other becomes homozygous for SNPs distal to the event.

To date we have analyzed 15,085 colonies from cells exposed to replication stress due to decreased expression of *POL1*. Of these, 38 colonies appeared as red/white sectors. Seven of the sectored colonies resulted from RCOs on the right arm of chromosome III, a frequency of 5.84×10^{-6} RCO/kb. This is a 10-fold increase over the spontaneous RCO frequency of 3.36×10^{-7}/kb observed by Barbara and Petes[10] on the left arm of yeast chromosome V (there are no fragile sites in this region). Six of the RCOs we obtained are located at fragile site FS2, and one is centromere-distal to FS2. Thus, our initial results support the hypothesis that instability at fragile sites stimulates mitotic recombination resulting in LOH. The remaining 31 sectored colonies all result from break-induced replication events. It is interesting that break-induced replication events are more frequently observed than RCOs. Because break-induced replication is thought to be favored when only one side of a double-strand break is available for repair, we hypothesize that one-ended breaks are frequent in our system as a result of the particular type of replication stress used. A low level of polymerase alpha is expected to cause replication fork stalling, and collapse of a stalled fork would produce a one-end double-strand break.

Discussion

CFSs are a normal part of human chromosome structure, yet instability at these loci under conditions of replication stress may contribute to genomic changes that are involved in both tumor initiation and progression. Because CFSs lack a defining linear base sequence motif that might point to a clear mechanism for how breaks form in these regions, extensive research has been undertaken to investigate multiple models that have been proposed to explain these sites of genomic instability. All of these models account for the fact that one characteristic common to all CFSs studied at the molecular level is late completion of DNA replication. These models include replication fork stalling in regions of DNA sequence with a high tendency to form intrastrand secondary structures; replication dynamics at early/late replication transition zones; origin initiation paucity; and transcription–replication collisions.[9,17–21] The first two models imply that certain regions within CFS loci are expected to be hotspots for breaks. Using as a model system yeast with a YAC carrying human fragile site DNA, our initial results reported here suggest that DNA flexibility peaks (defined as regions identified with a 100 bp sliding window to have an average twist angle above $13.7°$), do not appear to be favored break sites. We plan to conduct further fine-structure analysis to identify additional breakpoints, as well as to examine other DNA sequences that tend to form secondary structures. Further research will enable us to determine which of these four models is most appropriate, or whether CFSs are not a unified group but instead can be subdivided based on the model that best explains the instability of CFSs with distinct properties.

The connections between CFS and genomic changes in cancer have been studied extensively. Research in this area has been focused primarily on describing the changes present at CFS in cancer cells and on determining whether genes in or near CFS are tumor suppressors or oncogenes.[4–12] Instability at CFS clearly results in deletion, ampli-

fication, and translocation in tumors. The possible contribution of CFS to LOH by stimulating mitotic recombination, as suggested by the work reported here, calls for further investigation. Our yeast model system facilitates detailed study of mitotic recombination events between homologous chromosomes when subjected to replication stress. Cancer cells too are likely to be dividing under conditions of replication stress, and many cancer treatments cause replication stress. Thus, it is important to understand the extent of, and conditions that affect, fragile site stimulation of LOH in tumors, which contribute to tumor evolution and progression.

Acknowledgments

The authors wish to thank members of the Casper lab for assistance with experiments, and attendees of the DIMACS conference "Effects of Genome Structure and Sequence on the Generation of Variation and Evolution" for helpful discussions. A.M.C. is supported by NIGMS Grant R15GM093929.

Conflicts of interest

The authors declare no conflicts of interest.

References

1. Glover, T.W. *et al.* 1984. DNA polymerase alpha inhibition by aphidicolin induces gaps and breaks at common fragile sites in human chromosomes. *Hum. Genet.* **67:** 136–142.
2. Moore, J.M., H. Wimberly, P.C. Thornton, *et al.* Gross chromosomal rearrangement mediated by DNA replication in stressed cells: evidence from *Escherichia coli*. *Ann. N.Y. Acad. Sci.* **1267:** 103–109. This volume.
3. Burrow, A.A. *et al.* 2009. Over half of breakpoints in gene pairs involved in cancer-specific recurrent translocations are mapped to human chromosomal fragile sites. *BMC Genomics* **10:** 59.
4. McAvoy, S. *et al.* 2007. Non-random inactivation of large common fragile site genes in different cancers. *Cytogenet. Genome. Res.* **118:** 260–269.
5. Drusco, A. *et al.* 2011. Common fragile site tumor suppressor genes and corresponding mouse models of cancer. *J. Biomed. Biotechnol.* **2011:** 984505.
6. Blumrich, A. *et al.* 2011. The FRA2C common fragile site maps to the borders of MYCN amplicons in neuroblastoma and is associated with gross chromosomal rearrangements in different cancers. *Hum. Mol. Genet.* **20:** 1488–1501.
7. Pelliccia, F., N. Bosco & A. Rocchi. 2010. Breakages at common fragile sites set boundaries of amplified regions in two leukemia cell lines K562—Molecular characterization of FRA2H and localization of a new CFS FRA2S. *Cancer Lett.* **299:** 37–44.
8. Miller, C.T. *et al.* 2006. Genomic amplification of MET with boundaries within fragile site FRA7G and upregulation of

MET pathways in esophageal adenocarcinoma. *Oncogene* **25:** 409–418.

9. Durkin, S.G. & T.W. Glover. 2007. Chromosome fragile sites. *Annu. Rev. Genet.* **41:** 169–192.

10. Barbera, M.A. & T.D. Petes. 2006. Selection and analysis of spontaneous reciprocal mitotic cross-overs in Saccharomyces cerevisiae. *Proc. Natl. Acad. Sci. USA* **103:** 12819–12824.

11. Lee, P.S. *et al.* 2009. A fine-structure map of spontaneous mitotic crossovers in the yeast Saccharomyces cerevisiae. *PLoS Genet.* **5:** e1000410.

12. Dillon, L.W., A.A. Burrow & Y.H. Wang. 2010. DNA instability at chromosomal fragile sites in cancer. *Curr. Genomics* **11:** 326–337.

13. Pelliccia, F. *et al.* 2008. Replication timing of two human common fragile sites: FRA1H and FRA2G. *Cytogenet. Genome. Res.* **121:** 196–200.

14. Le Beau, M.M. *et al.* 1998. Replication of a common fragile site, FRA3B, occurs late in S phase and is delayed further upon induction: implications for the mechanism of fragile site induction. *Hum. Mol. Genet.* **7:** 755–761.

15. Hellman, A. *et al.* 2000. Replication delay along FRA7H, a common fragile site on human chromosome 7, leads to chromosomal instability. *Mol. Cell Biol.* **20:** 4420–4427.

16. Palakodeti, A. *et al.* 2004. The role of late/slow replication of the FRA16D in common fragile site induction. *Gen. Chromo. Cancer* **39:** 71–76.

17. Chan, K.L. *et al.* 2009. Replication stress induces sister-chromatid bridging at fragile site loci in mitosis. *Nat. Cell Biol.* **11:** 753–760.

18. El Achkar, E. *et al.* 2005. Premature condensation induces breaks at the interface of early and late replicating chromosome bands bearing common fragile sites. *Proc. Natl. Acad. Sci. USA* **102:** 18069–18074.

19. Palumbo, E. *et al.* 2010. Replication dynamics at common fragile site FRA6E. *Chromosoma* **119:** 575–587.

20. Letessier, A. *et al.* 2011. Cell-type-specific replication initiation programs set fragility of the FRA3B fragile site. *Nature* **470:** 120–123.

21. Helmrich, A., M. Ballarino & L. Tora. 2011. Collisions between replication and transcription complexes cause common fragile site instability at the longest human genes. *Mol. Cell* **44:** 966–977.

22. Mishmar, D. *et al.* 1998. Molecular characterization of a common fragile site (FRA7H) on human chromosome 7 by the cloning of a simian virus 40 integration site. *Proc. Natl. Acad. Sci. USA* **95:** 8141–8146.

23. Zlotorynski, E. *et al.* 2003. Molecular basis for expression of common and rare fragile sites. *Mol. Cell Biol.* **23:** 7143–7151.

24. Lemoine, F.J. *et al.* 2005. Chromosomal translocations in yeast induced by low levels of DNA polymerase a model for chromosome fragile sites. *Cell* **120:** 587–598.

25. Wilke, C.M. *et al.* 1996. FRA3B extends over a broad region and contains a spontaneous HPV16 integration site: direct evidence for the coincidence of viral integration sites and fragile sites. *Hum. Mol. Genet.* **5:** 187–195.

26. Zimonjic, D.B. *et al.* 1997. Positions of chromosome 3p14.2 fragile sites (FRA3B) within the FHIT gene. *Cancer Res.* **57:** 1166–1170.

27. Albertsen, H.M. *et al.* 1990. Construction and characterization of a yeast artificial chromosome library containing seven haploid human genome equivalents. *Proc. Natl. Acad. Sci. USA* **87:** 4256–4260.

28. Dausset, J. *et al.* 1992. The CEPH YAC library. *Behring Inst. Mitt.* **91:** 13–20.

29. Zhou, J. *et al.* 2000. Pif1p helicase, a catalytic inhibitor of telomerase in yeast. *Science* **289:** 771–774.

30. Chung, W.H. *et al.* 2010. Defective resection at DNA double-strand breaks leads to de novo telomere formation and enhances gene targeting. *PLoS Genet.* **6:** e1000948.

31. Marrero, V.A. & L.S. Symington. 2010. Extensive DNA end processing by exo1 and sgs1 inhibits break-induced replication. *PLoS Genet.* **6:** e1001007.

32. Lydeard, J.R. *et al.* 2010. Sgs1 and exo1 redundantly inhibit break-induced replication and de novo telomere addition at broken chromosome ends. *PLoS Genet.* **6:** e1000973.

33. Berger, A.H., A.G. Knudson & P.P. Pandolfi. 2011. A continuum model for tumour suppression. *Nature* **476:** 163–169.

34. Cole, F., S. Keeney & M. Jasin. 2012. Preaching about the converted: how meiotic gene conversion influences genomic diversity. *Ann. N.Y. Acad. Sci.* **1267:** 95–102. This volume.

Ann. N.Y. Acad. Sci. ISSN 0077-8923

ANNALS OF THE NEW YORK ACADEMY OF SCIENCES

Issue: *Effects of Genome Structure and Sequence on Variation and Evolution*

The genome: an isochore ensemble and its evolution

Giorgio Bernardi

Department of Biology, University Rome, Rome, Italy

Address for correspondence: Georgio Bernardi, Department of Biology, University Rome 3, Viale Marconi 446, Rome, 00146, Italy. gbernardi@uniroma3.it

The genomes of eukaryotes are mosaics of isochores. These are long DNA stretches that are fairly homogeneous in base composition and that belong to a small number of families characterized by different ratios of GC to AT and different short-sequence patterns (i.e., different DNA structures that interact with different proteins). This genome organization led to two discoveries: (1) the genomic code, which refers to two correlations, that of the composition of coding and contiguous noncoding sequences, and that of coding sequences and the structural properties of the encoded proteins; and (2) the genome phenotypes, which correspond to the patterns of isochore families in the genomes. These patterns indicate that genome evolution may proceed either according to a conservative mode or to a transitional (isochore shifting) mode, apparently depending upon whether the environment is constant or shifting. According to the neoselectionist theory, natural selection is responsible for both modes.

Keywords: genome organization; genome evolution; isochors; eukaryotes

The compositional strategy

Well before genomic sequences were available, we developed a compositional strategy to understand the organization of the eukaryotic genome.[1] The compositional strategy is based on the most elementary property of DNA, base composition. Originally, this strategy relied on the fractionation and characterization of DNA fragments (<30 to >300 kb in size) from eukaryotic genomes by preparative ultracentrifugation in density gradients of Cs_2SO_4 run in the presence of sequence-specific ligands, such as Ag^+ ions. The compositional strategy fractionated DNA fragments according to the density of short nucleotide sequences that bound the ligands. Because short sequences determine the structure of DNA and the interactions of DNA with proteins, such as histones to build nucleosomes, and transcription factors to interact with *cis*-regulatory sequences, the compositional approach actually led to a fractionation of the genome on the basis not only of its structure but also of its function. Needless to say, the compositional approach was applied to genome sequences as soon as they became available.

The function and structure of isochores

The compositional strategy led to the discovery that standard calf thymus DNA preparations (neglecting satellites) consisted of 10–50 kb fragments that belonged to a small number of families characterized by widely different GC levels[2] and by different short-sequence patterns.[3] This observation was shown to be valid for the genomes of all vertebrates and invertebrates explored.[4] In fact, we found that the DNA fragments were derived from much larger DNA stretches, originally defined as genome regions >300 kb in size, that were fairly homogeneous in base composition and that could be grouped into the same families as the DNA fragments derived from them.[5] We later called these large regions *isochores* for their similar landscapes.[6]

This isochore mosaic structure of the vertebrate genome became much easier to investigate once entire chromosomal DNA sequences became available, and thus could be analyzed computationally rather than by the use of density gradients. Indeed, if one scans a chromosome, for example, human chromosome 21, with nonoverlapping, fixed-length windows of 100 kb, one can observe 340 window-size

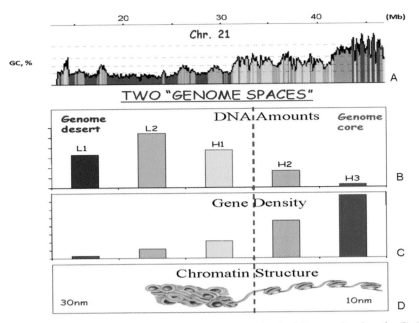

Figure 1. (A) The compositional profile of human chromosome 21 (comprising the full range of isochore families) as seen through nonoverlapping fixed windows of 100 kb.[7] (B) The histogram shows the isochores as pooled in bins of 1% GC. Isochore families are represented by different colors. The Gaussian profile shows the distribution of the five isochore families L1 to H3 from GC-poor to GC-rich, the genome phenotype.[7] (C) Gene densities of the isochore families belong in two genome spaces: a gene-rich genome core and a gene-poor genome desert. (Modified from Costantini *et al.*).[7] (D) Open and closed chromatin in the two genome spaces.[8]

segments that are characterized by fairly homogeneous GC levels over regions extending up to several megabases (Fig. 1A). When isochores from the whole genome are put in bins according to GC levels, isochore families become evident (Fig. 1B).

As first proposed by Cuny *et al.*,[6] isochores are responsible for the chromosomal bands originally observed by Caspersson *et al.*,[9] GC-poor and GC-rich isochores being predominant in the quinacrine (or Giemsa) positive and negative bands, respectively. However, isochores are not just important structures of chromosomes; they are also associated with the distribution of genes (Fig. 1C), with chromatin structure (see Fig. 1D), as well as with the timing of replication[10] and the probability of recombination.[11] In fact, several of these genome properties can independently partition the human genome into two main structural, functional, and evolutionary genome spaces, which we have termed the *genome core* and the *genome desert*, comprising the GC-rich and GC-poor isochores, respectively.

The isochore structure of the genome allowed us to discover[12,13] several compositional correlations, collectively called the *genomic code*.[1] The compositional correlations hold between coding and con-

tiguous noncoding sequences, as well as between coding sequences and the structural properties of the encoded proteins.[1,13] The genomic code implies that comprehensive rules of genome organization exist, as revealed by the existence of a genomic code; the genomic code also leads to the view of the genome as an integrated ensemble, rather than a string of independently mutating and selected nucleotides.

As already mentioned, we observed what we have termed genome phenotypes that correspond to the patterns of isochore families in the genome (see Fig. 1B). Comparisons of isochore patterns among vertebrates indicate that genome evolution may proceed according to a conservative mode (with respect to isochore composition and its associated biological properties) or to a transitional (or shifting) mode, and has opened a new way to investigate genome evolution.

Genome evolution

"Most of the familiar features of living organisms show clear signs of adaptation of structure to function. There is overwhelming evidence that this is the outcome of evolution by natural selection."[14] Whether the same applies to the genomes of vertebrates (and other eukaryotes) with their overwhelming amount of nonprotein-coding DNA was, however, an open question, which we approached by investigating the evolution of genome phenotypes.

A crucial preliminary remark is that changes in the organismal phenotype depend much more on changes in regulatory sequences than on changes in protein coding genes.[15] In turn, we must consider that there is an effect of the compositional context of a genome region on chromatin structure and thus on regulation of gene expression.

To explain the conservative mode of genome evolution, the *neoselectionist theory* proposes[16] that local clustering of nucleotide substitutions (which are AT-biased), as well as insertions/deletions, duplications, and the GC-biased changes (which are due to biased gene conversion (gBGC))[17] can alter both chromatin structure (nucleosome positioning and density) and *cis*-regulatory sequences and, therefore, affect the expression of genes. These "critical changes" especially concern the genes located in GC-rich, CpG-rich, and gene-rich isochores, which have much more frequent AT-biased changes[18] and insertions/deletions.[19] In turn, the resulting structural effects on regional gene expression may lead to negative (purifying) selection of the carriers of the altered genome and of their progeny, thus preserving the original genome phenotype. In other words, the neoselectionist theory visualizes the conservative mode of evolution as a process in which apparently minor sequence changes can have dramatic effects on chromatin structure and *cis*-regulatory sequences, and therefore gene expression, resulting in negative selection of the carriers of such alterations and their progeny (except in the extremely rare case that the dramatic change has a positive effect).

In the other observed mode of genome evolution, the transitional or shifting mode, positive selection plays a role by favoring AT/GC changes in coding sequences, as well as regional AT/GC changes (such as those due to gBGC), insertions of GC-rich interspersed repeats and of GC-rich microsatellites that favor DNA stability and gene expression. Negative selection may also play a role, however.[16]

In conclusion, the existence of a genomic code points to an integrated ensemble view of the genome (in contrast with the simple operational view that has been predominant since the original definition

of the genome as the haploid chromosome set).[20] The isochore mosaic may be compositionally conserved by natural selection (essentially negative or purifying selection) because compositional changes (such as those due to insertions/deletions, for instance) cause local alterations of chromatin and of gene expression (as a consequence). On the other hand, the emergence of warm-blooded vertebrates favored compositional changes (mainly by positive selection) that led to DNA stability at their higher body temperatures.

Conflicts of interest

The author declares no conflicts of interest.

References

1. Bernardi, G. 2004, reprinted in 2005. Structural and evolutionary genomics. In *Natural Selection in Genome Evolution*. Elsevier. Amsterdam. This book is freely available at www.giorgiobernardi.it, a website from which all publications from Bernardi's laboratory can be downloaded.
2. Filipski, J., J.P. Thiery & G. Bernardi. 1973. An analysis of the bovine genome by Cs2 SO4 /Ag+ density gradient centrifugation. *J. Mol. Biol.* **80:** 177–197.
3. Arhondakis, S., F. Auletta & G. Bernardi. 2011. Isochores and the regulation of gene expression in the human genome. *Genome Biol. Evol.* **3:** 1080–1089.
4. Thiery, J.P., G. Macaya & G. Bernardi. 1976. An analysis of eukaryotic genomes by density gradient centrifugation. *J. Mol. Biol.* **108:** 219–235.
5. Macaya, G., J.P. Thiery & G. Bernardi. 1976. An approach to the organization of eukaryotic genomes at a macromolecular level. *J. Mol. Biol.* **108:** 237–254.
6. Cuny, G., P. Soriano, G. Macaya & G. Bernardi. 1981. The major components of the mouse and human genomes: preparation, basic properties and compositional heterogeneity. *Eur. J. Biochem.* **111:** 227–233.
7. Costantini, M., O. Clay, F. Auletta & G. Bernardi. 2006. An isochore map of human chromosomes. *Genome Res.* **16:** 536–541.
8. Di Filippo, M. & G. Bernardi. 2008. Mapping DNase I-hypersensitive sites on human isochores. *Gene.* **419:** 62–65.
9. Caspersson, T., S. Farber, G.E. Foley, *et al.* 1968. Chemical differentiation along metaphase chromosomes. *Exp. Cell. Res.* **49:** 219–222.
10. Costantini, M. & G. Bernardi. 2008. Replication timing, chromosomal bands and isochores. *Proc. Natl. Acad. Sci. USA* **105:** 3433–3437.
11. Fullerton, S.M., A.B. Carvalho & A.G. Clark. 2001. Local rates of recombination are positively correlated with GC content in the human genome. *Mol. Biol. Evol.* **18:** 1139–1142.
12. Bernardi, G., B. Olofsson, J. Filipski, *et al.* 1985. The mosaic genome of warm-blooded vertebrates. *Science* **228:** 953–958.
13. Bernardi, G. & G. Bernardi. 1986. Compositional constraints and genome evolution. *J. Mol. Evol.* **24:** 1–11.

14. Charlesworth, B. 2008. The origins of genomes—not by natural selection? *Curr. Biol.* **18:** 140–141.

15. King, M.C. & A.C. Wilson. 1975. Evolution at two levels in humans and chimpanzees. *Science* **188:** 107–116.

16. Bernardi, G. 2007. The neo-selectionist theory of genome evolution. *Proc. Natl. Acad. Sci. USA* **104:** 8385–8390.

17. Eyre-Walker, A. & L.D. Hurst. 2001. The evolution of isochores. *Nat. Rev. Genet.* **2:** 549–555.

18. Li, M.K., L. Gu, S.S. Chen, *et al.* 2008. Evolution of the isochore structure in the scale of chromosomes: insight from the mutation bias and fixation bias. *J. Evol. Biol.* **21:** 173–182.

19. Costantini, M. & G. Bernardi. 2009. Mapping insertions, deletions and SNPs on Venter's chromosomes. *PLoS One* **4:** e5972.

20. Winkler, H. 1920. *Verbreitung und Ursache der Parthenogenesis im Pflanzen und Tierreich.* Fischer. Jena.

Ann. N.Y. Acad. Sci. ISSN 0077-8923

ANNALS OF THE NEW YORK ACADEMY OF SCIENCES

Issue: *Effects of Genome Structure and Sequence on Variation and Evolution*

Multiple levels of meaning in DNA sequences, and one more

Edward N. Trifonov,[1,2] Zeev Volkovich,[3] and Zakharia M. Frenkel[1,3]

[1]Genome Diversity Center, Institute of Evolution, University of Haifa, Mount Carmel, Haifa, Israel. [2]Central European Institute of Technology and Faculty of Science, Masaryk University, Brno, Czech Republic. [3]Department of Software Engineering, ORT Braude College, Karmiel, Israel

Address for correspondence: Edward N. Trifonov, Genome Diversity Center, Institute of Evolution, University of Haifa, Mount Carmel, Haifa 31905, Israel. trifonov@research.haifa.ac.il

If we define a *genetic code* as a widespread DNA sequence pattern that carries a message with an impact on biology, then there are multiple genetic codes. Sequences involved in these codes overlap and, thus, both interact with and constrain each other, such as for the triplet code, the intron-splicing code, the code for amphipathic alpha helices, and the chromatin code. Nucleosomes preferentially are located at the ends of exons, thus protecting splice junctions, with the N9 positions of guanines of the GT and AG junctions oriented toward the histones. Analysis of protein-coding sequences reveals numerous traces of tandem repeats, apparently formed by triplet expansion, which in effect is a genome inflation "code". Our data are consistent with the hypothesis that expansion of simple tandem repetition of certain aggressive triplets has been a characteristic of life from its emergence. Such expanding triplets appear to be the major factor underlying observed codon usage biases.

Keywords: sequence codes; overlapping codes; gene splicing; nucleosome positioning; genome inflation code; codon usage

Introduction

The term *genetic code* is widely used to refer to the table of codons, in which nucleotide triplets in protein-coding sequences correspond to specific amino acids. Over the past few decades, our work has revealed a multiplicity of sequence codes that overlap,[1–3] a property that is enabled by their degeneracy. For example, in 1980, we suggested what turned out to be the first of many nontriplet codes—the chromatin code,[4,5] which recently has gained significant attention[6] (a "second" genetic code). Thus *Nature*[7] and the *New York Times*[8] gave due publicity to the concept that there is not just one code in the genomic sequences. Interestingly, the thought that the triplet code is not alone was again pointed out by a *Nature* paper,[9,10] describing yet another "second" genetic code, this time the gene-splicing code. In fact, about 10 different "second" genetic codes are known today.[3]

The chromatin code

Nucleosome positioning

Owing to the availability of large sequence ensembles and of databases of nucleosome DNA sequences, the chromatin code has finally been cracked within the last three years,[11–14] down to the finest details. The nucleosome positioning–sequence pattern derived allows for the mapping of nucleosomes onto DNA sequences with a single-base resolution[12] that has been verified by earlier X-ray diffraction data on crystals of nucleosomes formed on specific sequences.[15] This has allowed determination of which bases of the nucleosome DNA are exposed to the exterior of the nucleosome and which ones are hidden on the interface between DNA and histone octamer.

Relationship to splice junctions

This high-resolution nucleosome mapping was used in our recent study on protective positioning of

doi: 10.1111/j.1749-6632.2012.06589.x

intron splice sites in the nucleosomes[16] to examine the early observation[17] that the splice junctions are protected by the nucleosomes. It was found that the N9 atoms of the guanine residues of the AG and GT intron ends are preferentially located closest to the surface of the histone octamers, thus being least accessible to depurination attacks.[17] This study provides a good example of the overlap (and interaction) of two different codes—the chromatin code and the gene-splicing code.

Curved DNA and ampipathic helices

One spectacular example of the overlap of different codes is the motif CGGAAATTCCG, responsible for nucleosome positioning.[11] A very similar periodically repeating motif, GAAAATTTTC, corresponds to curved DNA;[18] in addition, another very similar motif, GGGAATTTCC, represents the consensus sequence for binding sites of the important transcription factor NF-κB.[19] Finally, the periodically repeating motif, GAAAATTTTC, contributes to sequences that encode amphipathic alpha helices in proteins.[20] There are, thus, several selective pressures that would serve to maintain this motif. To examine how widespread this motif in fact is, we used Shannon N-gram extension. This method involves taking the most common triplets, ranking them, and observing in what patterns they are found. For example, if the triplets ACG, CGT, and GTT are the most frequent in the list of 64, then the "extension" motif ACGTT would be frequent as well. It will be detected even when, because of mutations, the intact extension motif does not appear, leaving only its traces in form of the triplets. Indeed, using Shannon N-gram extension of the most frequent "words" of oligonucleotide vocabularies reveals that GAAAATTTTC is the most dominant hidden motif in both eukaryotes[21] and prokaryotes.[20] Thus, the nucleosome-positioning chromatin code is the major code in eukaryotes across the entire genome (i.e., protein coding and nonprotein coding), while in prokaryotes, even without nucleosomes, the same pattern is dominant and codes for amphipathic alpha helices. It may well serve in addition as a motif to bind histone-like proteins. For genome evolution, what we now view as the nucleosome-positioning pattern seems to have emerged first as a dominant DNA sequence pattern in prokaryotes.[20]

Repetitive DNA sequences

Rapid adaptation

Another code that seems to play an important role in genome adaptation during evolution is the modulation (fast adaptation) code described first in Ref. 1 and subsequently formulated by several research groups independently.[22–24] (This is discussed in Ref. 25 as a genomic protocol that facilitates rapid adaptation.) The length of simple polymorphic sequence repeats changes rapidly. The changes in the length (copy number) of the repeats are associated with a range of biological effects, including changes in the level of expression of nearby genes.[25,26] Such sequences spontaneously emerge in many locations in the genome and, where they provide an advantage by enabling rapid adaptation to changing environments, would be retained by selection (see further discussion of such indirect selection of repetitive sequences in Ref. 25).

Genome expansion

In our latest studies on the evolution of genome sequences, we came to realize that, in addition to facilitating rapid adaptation, tandemly repeating sequences provide a fundamental building material for evolving genomes. They can be considered loci of "genome inflation."[27] Our analysis suggests that much of the evolutionary history of genomes, dating back to the onset of life on our planet, involves the expansion of tandem triplet repeats,[28–30] followed by mutational changes, leading to eventual accommodation of all triplets as codons of the triplet code,[29] further triplet expansions within the evolving genomes, and mutational transformations of the repeats toward higher sequence complexity and functionality.[27,31] This ongoing process of local expansions of various tandem repeats and their subsequent accommodation seem to be major genome-forming processes.[27,31]

Effect on codon usage

This source of genome inflation sheds new light on the classical triplet code. One of the most enigmatic features of the code is uneven codon usage. Many attempts to rationalize this phenomenon provide only partial explanations.[32–34] Analysis of a very large database of nonredundant prokaryotic and eukaryotic protein-coding sequences, of the size of 5×10^9 codons,[31] has shown that the major factor that shapes codon usage seems to be triplet

Ala	GCC	110	465	Arg	CGC	70	177	Arg	AGA	55	62
	GCA	94	195		CGU	46	45		AGG	29	22
	GCU	93	245		CGG	41	86				
	GCG	88	386		CGA	33	39				
Asn	AAU	121	523	Asp	GAU	148	359	Cys	UGC	31.9	18
	AAC	85	170		GAC	107	236		UGU	31.5	7
Gln	CAA	88	269	Glu	GAA	163	584	Gly	GGC	107	500
	CAG	87	459		GAG	122	367		GGU	92	229
									GGA	87	135
									GGG	56	17
His	CAU	58	62	Ile	AUU	128	151	Leu	UUA	91	127
	CAC	49	61		AUC	100	107		UUG	73	30
					AUA	70	63				
Leu	CUG	108	375	Lys	AAA	158	403	Met	AUG	109	117
	CUU	75	43		AAG	104	277				
	CUC	70	59								
	CUA	40	8								
Phe	UUU	112	68	Pro	CCA	62	89	Ser	UCU	63	81
	UUC	82	85		CCG	59	169		UCA	62	90
					CCU	58	59		UCC	50	67
					CCC	50	11		UCG	44	54
Ser	AGC	59	147	Thr	ACC	76	138	Trp	UGG	60	22
	AGU	53	36		ACA	71	126				
					ACU	65	45				
					ACG	51	59				
Tyr	UAU	86	68	Val	GUG	91	187				
	UAC	61	41		GUU	88	92				
					GUC	74	103				
					GUA	61	23				

Figure 1. Codon usage in nonrepetitive parts of mRNA. Topmost codons of every repertoire matching to higher frequency of the repeats are shown in bold. Left columns of numbers correspond to codon counts ($\times 10^{-6}$); right columns correspond to repeat counts ($\times 10^{-3}$).

expansions. Those triplets that are more "aggressive" (i.e., more repetitive) are also dominant in the respective codon repertoires. This is illustrated in Figure 1, where the codons are grouped in 21 repertoires. In 17 of 21 repertoires, the codons of highest usage are also the triplets of highest repetitiveness.

Nonprotein-coding regions

We also examined the possibility that triplet expansions are an evolutionary force in nonprotein-coding sequences. For this purpose, a simple algorithm for selection of DNA sequence fragments that retain evidence of earlier triplet expansion events was developed:

(1) For each of 64 triplets, all regions of possible expansion were defined as regions of 12 bases or longer that contain only the corresponding triplet or its single-base mutation derivatives.

(2) All these regions were colored, and the proportion of regions that were colored was calculated. This result was compared with the

same analysis but using shuffled sequences (the shuffling was carried out on a large number of sequences keeping the overall composition of triplets the same).

The proportion of the colored sequences in the coding-sequence database was 22.3% in natural and 13.8% in shuffled cDNA sequences, respectively. When this algorithm was applied for investigation of potential ancient triplet expansion events in the human genome (18th chromosome, about 25×10^6 triplets), which is ~96% noncoding, the result was very similar to the one described above: 20.3% and 13.9%, respectively. That is, for all sequences, protein coding and nonprotein coding alike, sequences that seem to have evolved largely by local events of triplet expansion, followed by mutational changes in the repeats, are dramatically overrepresented.

Using the above technique, we were able to compare the expansion efficiencies of various triplets in the protein-coding and nonprotein-coding regions (in preparation). For example, in nonprotein-coding sequences, most expandable triplets are AAA and TTT, whereas in coding sequences their expansion is substantially less. GCC triplets, on the other hand, are an order of magnitude more expandable in the protein-coding regions versus noncoding sequences. These and other differences reflect different selection pressures on the coding- and noncoding-sequences (such as pressure to avoid frameshifts caused by single base changes in coding regions) and invite further studies of the genome inflation phenomenon.

Conclusion

Computational analysis reveals multiple levels of meaning in genomic sequences, much of which have been conserved over vast evolutionary timescales. The sequence material for development of new sequence patterns (and meanings) appears to be largely provided by spontaneously expanding tandem repeats, inflating the evolving genomes.

Conflicts of interest

The authors declare no conflicts of interest.

References

1. Trifonov, E.N. 1989. The multiple codes of nucleotide sequences. *Bull. Math. Biol.* **51:** 417–432.

2. Trifonov, E.N. 1999. Sequence codes. In *Encyclopedia of Molecular Biology*. T. E. Creighton, Ed.: 2324–2326. John Wiley & Sons, Inc. New York.

3. Trifonov, E.N. 2011. Thirty years of multiple sequence codes. *Genom. Proteom. Bioinform.* **9:** 1–6.

4. Trifonov, E.N. 1980. Sequence-dependent deformational anisotropy of chromatin DNA. *Nucl. Acids Res.* **8:** 4041–4053.

5. Trifonov, E.N. 1981. Structure of DNA in chromatin. In *International Cell Biology 1980–1981*. H. Schweiger, Ed.: 128–138. Springer-Verlag. Berlin.

6. Segal, E., Y. L. Fondufe-Mittendorf, A. Chen, *et al.* 2006. A genomic code for nucleosome positioning. *Nature* **442:** 772–778.

7. Pearson, H. 2006. Genetic information: codes and enigmas. *Nature* **444:** 259–261.

8. Wade, N. 2006. Scientists Say They've Found a Code Beyond Genetics in DNA. *New York Times*. July 25.

9. Barash, Y., J. Calarco, W. Gao, *et al.* 2010. Deciphering the splicing code. *Nature* **465:** 53–59.

10. Tejedor, J.R. & J. Valcárcel. 2010. Gene regulation: breaking the second genetic code. *Nature* **465:** 45–46.

11. Gabdank, I., D. Barash & E.N. Trifonov. 2009. Nucleosome DNA bendability matrix (C. elegans). *J. Biomol. Str. Dyn.* **26:** 403–412.

12. Gabdank, I., D. Barash & E.N. Trifonov. 2010. Single-base resolution nucleosome mapping on DNA sequences. *J. Biomol. Struct. Dyn.* **28:** 107–121.

13. Trifonov, E.N. 2011. Cracking the chromatin code: precise rule of nucleosome positioning. *Physics of Life Reviews* **8:** 39–50.

14. Frenkel, Z.M., T. Bettecken & E.N. Trifonov. 2011. Nucleosome DNA sequence structure of isochores. *BMC Genomics* **12:** 203.

15. Richmond, T.J. & C.A. Davey. 2003. The structure of DNA in the nucleosome core. *Nature* **423:** 145–150.

16. Hapala, J. & E.N. Trifonov. 2011. High resolution positioning of intron ends on the nucleosomes. *Gene.* **489:** 6–10.

17. Denisov, D.A., E.S. Shpigelman & E.N. Trifonov. 1997. Protective nucleosome centering at splice sites as suggested by sequence-directed mapping of the nucleosomes. *Gene.* **205:** 145–149.

18. Hagerman, P.J. 1986. Sequence-directed curvature of DNA. *Nature* **321:** 449–450.

19. JASPAR, database of transcription factor binding sites.

20. Rapoport, A.E. & E.N. Trifonov. 2011. "Anticipated" nucleosome positioning pattern in prokaryotes. *Gene.* **488:** 41–45.

21. Rapoport, A.E., Z.M. Frenkel & E.N. Trifonov. 2011. Nucleosome positioning pattern derived from oligonucleotide compositions of genomic sequences. *J. Biomol. Struct. Dyn.* **28:** 567–574.

22. Holliday, R. 1991. Quantitative genetic variation and developmental clocks. *J. Theor. Biol.* **151:** 351–358.

23. King, D.G. 1994. Triple repeat DNA as a highly mutable regulatory mechanism. *Science* **263:** 595–596.

24. Künzler, P., K. Matsuo & W. Schaffner. 1995. Pathological, physiological, and evolutionary aspects of short unstable DNA repeats in the human genome. *Biol. Chem. Hoppe-Seyler* **376:** 201–211.

25. King, D. 2012. Indirect selection of implicit mutation protocols. *Ann. N.Y. Acad. Sci.* **1267:** 45–52. This volume.

26. Bayliss, D.D. & M.E. Palmer. 2012. Evolution of simple sequence repeat–mediated phase variation in bacterial genomes. *Ann. N.Y. Acad. Sci.* **1267:** 39–44. This volume.

27. Koren, Z. & E.N. Trifonov. 2011. Role of everlasting triplet expansions in protein evolution. *J. Mol. Evol.* **72:** 232–239.

28. Trifonov, E.N. 2000. Consensus temporal order of amino acids and evolution of the triplet code. *Gene.* **261:** 139–151.

29. Trifonov, E.N. 2004. The triplet code from first principles. *J. Biomolec. Str. Dyn.* **22:** 1–11.

30. Trifonov, E.N. 2009. Origin of the genetic code and of the earliest oligopeptides. *Res. Microbiol.* **160:** 481–486.

31. Frenkel, Z.M. & E.N. Trifonov. 2012. Origin and evolution of genes and genomes. Crucial role of triplet expansions. *J. Biomol. Str. Dyn.* **30:** 201–210.

32. Hershberg, R. & D. A. Petrov. 2008. Selection on Codon Bias. *Ann. Rev. Genetics* **42:** 287–299.

33. Plotkin, J.B. & G. Kudla. 2011. Synonymous but not the same: the causes and consequences of codon bias. *Nat. Rev. Gen.* **12:** 32–42.

34. Sauna, Z.E. & C. Kimchi-Sarfaty. 2011. Understanding the contribution of synonymous mutations to human disease. *Nat. Rev. Gen.* **12:** 683–691.

Ann. N.Y. Acad. Sci. ISSN 0077-8923

ANNALS OF THE NEW YORK ACADEMY OF SCIENCES
Issue: *Effects of Genome Structure and Sequence on Variation and Evolution*

Evolution of simple sequence repeat–mediated phase variation in bacterial genomes

Christopher D. Bayliss[1] and Michael E. Palmer[2]

[1]Department of Genetics, University of Leicester, Leicester, United Kingdom. [2]Department of Biology, Stanford University, Stanford, California

Address for correspondence: Christopher D. Bayliss, Department of Genetics, University of Leicester, University Road, Leicester, LE1 7RH, United Kingdom. cdb12@le.ac.uk

Mutability as mechanism for rapid adaptation to environmental challenge is an alluringly simple concept whose apotheosis is realized in simple sequence repeats (SSR). Bacterial genomes of several species contain SSRs with a proven role in adaptation to environmental fluctuations. SSRs are hypermutable and generate reversible mutations in localized regions of bacterial genomes, leading to phase variable ON/OFF switches in gene expression. The application of genetic, bioinformatic, and mathematical/computational modeling approaches are revolutionizing our current understanding of how genomic molecular forces and environmental factors influence SSR-mediated adaptation and led to evolution of this mechanism of localized hypermutation in bacterial genomes.

Keywords: phase variation; simple sequence repeat; bacteria; pathogen

Introduction

Single-celled organisms living in contact with other adaptable and evolving organisms are subject to unpredictable and fluctuating selective pressures. Survival of these pressures requires high levels of genetic or epigenetic variation, particularly in those genes whose products directly influence interactions with the environment, such as the surface molecules responsible for binding to other organisms and surfaces. A number of mechanisms have evolved in bacterial genomes for focusing variation to specific regions of bacterial genomes, including differential methylation of promoter elements and site-specific recombination.[1–3] Variation occurs stochastically at high frequencies and is often reversible, resulting in the phenomenon of phase variation (PV), that is, ON/OFF switches in expression of surface molecules. PV is of major importance for our understanding of bacterial virulence, as key virulence determinants of a wide range of bacterial pathogens are subject to switches in gene expression by this mode.

Simple sequence repeats (SRR), or microsatellites, are a major mechanism of PV. These highly mutable sequences are present in many bacterial genes, with the genomes of major pathogens, such as *Neisseria meningitidis, Haemophilus influenzae, Helicobacter pylori,* and *Campylobacter jejuni,* containing 10–40 genes that are subject to SSR-mediated PV (see Ref. 2). Reversible alterations in the lengths of SSR repeat tracts occur on either the template or nascent strand during DNA replication, probably due to slipped-strand mispairing.[4,5] Mutations in SSRs within the coding region shift the reading frame, resulting in two (ON/OFF) or possibly three levels of gene expression; for example, a change from G8 to G7 or G9 in a poly(G) repeat tract located in the middle of *cj1139* results in an ON-to-OFF switch in expression of this gene in *C. jejuni* strain 11168.[6] Mutations in promoter-located SSRs generate more subtle variations in protein expression levels due to alterations in the relative spacing and positional orientation of promoter elements on the DNA helix; for example, a change from TA10 to TA9, in a TA dinucleotide repeat tract located between the −10 and −35 elements of the *hif* promoter, significantly reduces expression of pili in *H. influenzae,* whereas a change from TA10 to TA11 produces only a moderate change.[7] The low coding capacity, simplicity,

doi: 10.1111/j.1749-6632.2012.06584.x

and dynamic nature of SSRs mark repeat–mediated PV as being a highly evolvable mechanism for generation of genetic variation and one that is tractable to investigation by a range of techniques. Major advances in recent years have uncovered novel ideas regarding the mutational mechanisms and selective pressures acting on these repeats, with strong indications of the range of the processes driving evolution of this form of dynamic DNA being a mechanism of bacterial adaptability. See Barry[8] for a discussion of surface antigen variation in trypanosomes.

Molecular drivers of repeat-mediated PV rate

The mutability of SSRs in phase variable loci is determined by a combination of environmental (selection/epoch length), population (bottlenecks/genetic drift), or molecular drivers (Fig. 1). The molecular drivers are forces intrinsic to a specific locus or genome, such as the *trans*-acting factors and *cis*-acting sequences that determine the rate and relative proportions of insertions and deletions of repeats.[9] These drivers will affect evolution of these tracts; thus, an increase in the number of insertions will drive SSR tracts toward longer lengths, while a deletion bias will shorten these tracts. These molecular forces may vary between strains, as a function of tract length and from temporal fluctuations caused by changes in the environmental conditions. For example, environmental stress could alter the levels of *trans*-acting factors, leading to a "slippery cycle" that is characterized by more rapid changes in current tracts. Analysis of the molecular drivers is tractable by both experimental and bioinformatic approaches, as outlined in the following sections, and is critical for understanding SSR-mediated PV, as it provides measurements of the underlying mutational parameters.

The role of DNA replication, mismatch repair, and tract length

Using biochemical assays and reporter constructs, the replicative DNA machinery has been shown to be a major determinant of SSR mutability.[10–14] This machinery sets the uncorrected level of mutations in these tracts, but also influences tract evolution due to differential effects of leading- and lagging-strand synthesis on specific repeat sequence types and on tracts of differing lengths. As perturbation of DNA replication will reduce fitness, this machin-

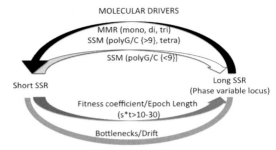

Figure 1. Evolution of repeat-mediated phase variable loci. Long SSR tracts are highly mutable and mediate high frequency, reversible switches in gene expression. These tracts evolve from shorter SSRs, due to a combination of molecular and environmental/population forces. The molecular forces are intrinsic to the genome of the bacterial species, although they can also be subject to alteration due to external selective forces. The environmental forces are the product of the fitness coefficient (*s*, strength of selection) and frequency of selection (*t*, epoch length) acting on the product of the phase variable loci and exerting secondary selection for mutability. The influence of the population drivers are poorly understood but include short time-scale bottlenecks (selective and nonselective), longer timescale population bottlenecks (genetic drift), and concerted effects of multiple phase variable genes. SSM, slipped-strand mispairing.

ery may have only a minimal effect on differences in PV rates between strains. In contrast, loss- and gain-of-mismatch repair (MMR) occurs readily in bacterial populations and has a major influence on mutability, increasing switching rates up to 1,000-fold.[15] However, MMR acts only on tracts consisting of repeating units of <4 nucleotides.[10,16] Another major factor controlling strain-to-strain variations in PV rates is tract length, as repeat number is directly correlated with mutability.[15,17] Tract length will be strongly constrained for tracts present in the core promoter, but more flexible when present in the reading frame, and hence it will make a greater contribution to the localized evolution of PV rates.

Mutational pattern as a determinant of repeat-tract length variation

In reporter-based studies performed *in vitro*, and hence in the absence of an environmental selective pressure, a twofold excess of deletions over insertions was observed for 5′AGTC repeats of 17–38 units in *H. influenzae*.[17] This shortening of the tracts is presumably counteracted by secondary selection for variation generated by mutability in the phase variable loci. Recently, examination of

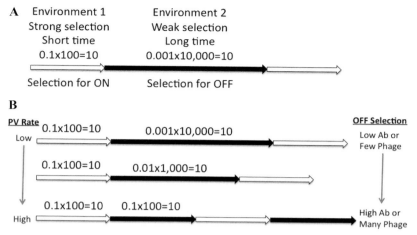

Figure 2. Influence of environmental factors on evolution of SSR-mediated phase variation. Modeling indicates that SSR-mediated PV evolves when the product of the fitness benefits or selection strength (s), multiplied by the length of the epoch in generations (t), is greater than 10 or 30 for the two expression states (see text and Palmer *et al.*). Panel A shows one example of how this periodic selection could be exerted on the ON and OFF states of a phase variable loci. Modeling also indicates that the actual switching rate is determined by the combined lengths of the two epochs ($1/T_{avg}$, where T_{avg} is the mean time spent in environments 1 and 2). Panel B indicates how a heightened selection for the OFF state, due to increasing amounts of antigenic-specific antibodies (Ab) or bacteriophages, results in a higher PV rate. In this case T_{avg}, and hence the favored PV rate, increases from a low, 2×10^{-4}, to a high, 1×10^{-2}, rate.

mononucleotide poly(G) tracts in phase variable loci of *C. jejuni* indicated a propensity for insertions in tracts of 8 and 9 units but deletions in tracts of 11 units.[6] As the majority of *C. jejuni* phase variable genes contain poly(G) tracts of 9 or 10 repeat units, the mutational pattern suggests that molecular drivers rather than selection are the main determinant of tract length.

Analysis of SSR mutability in whole genomes

SSRs are readily accessible to bioinformatic analyses, and two recent articles have analyzed "short" SSRs that provide the substrates for evolution of the "long" SSRs associated with PV. Lin and Kussell[18] found that SSRs were present at higher than expected densities in several proteobacteria, consistent with positive selection, and exhibited a biased distribution, with a higher prevalence in the N- and C-termini of proteins, suggesting protein structural constraints on SSR evolution. SSR mutability was affected by repeat sequence type and tract length across a diverse set of bacterial species. Notably, mono- and dinucleotide repeat tracts exhibited nonlinear increases in mutability as lengths exceeded 6–10 bp, raising the possibility of a threshold above which selection can act to further increase

mutability, leading to evolution of a phase variable locus.

In studies centered on mononucleotide tracts of <8 nucleotides, Kumar and Nagarajaram[19] detected a propensity for deletions in repeat tracts in multiple MMR-proficient species (e.g., *H. influenzae*), but a balance of insertions and deletions in species lacking MMR (e.g., *C. jejuni*), possibly indicative of efficient repair of insertions in the nascent strand by MMR-proficient species. Thus, as MMR reduces insertions and, hence, tract expansion, evolution of long mononucleotide tracts in MMR-proficient species may require strong selection for mutability and/or significant periods of evolution in MMR-deficient strains. Alternatively, mononucleotide repeat–mediated PV may rapidly evolve in MMR-deficient species, in combination with weak selection for mutability.

The importance of linkage and population size for repeat-mediated PV

The major force driving evolution of repetitive DNA–driven PV is secondary selection for mutability at each specific locus. The primary selection is for a change in the phenotype controlled by the specific locus, which simultaneously selects for the increase in mutability that enabled the appearance of that phenotype (see King[20]). Close linkage (when SSRs are present within the same gene) between the mechanism of mutability and the phenotype-encoding locus ensures that phenotypic variation is coinherited with the mutable mechanism.

PV is a population-based adaptive strategy. Adaptation involves outgrowth of a "fit" variant generated before encountering a change in the selective pressure (hence the term *contingency loci*[21]). The size of the population required for adaptation is determined by the switching rate, such that populations of 10^3–10^4 cells will usually contain at least one phase variant for most phase variable loci.

Computer modeling of the selective forces favoring evolution of phase variable loci

For phase variable genes, particularly those subject to ON and OFF switching, it is presumed that the switching mechanism has evolved by selection for adaptation to two recurring environments with opposite constraints. In this case, the selective force acting on these loci will consist of two components—the strength of the fitness benefit conferred by a change in phenotype for each environment and the frequency of the change between the two environments. Assessment of these forces is amenable to simulation in mathematical and computer models (Fig. 2).

Evolution of repeat-mediated PV was assessed in a model incorporating a locus in which the mutation rate was allowed to evolve (Palmer *et al.*, in preparation), and whose mutability and mutational patterns were derived from experimental data for tetranucleotide repeats in *H. influenzae* (see above). In the absence of selection, SSR tracts evolved to the shortest length (17 repeats in this model) due to a preponderance of deletions over insertions. Heightened PV rates and tracts of >17 repeats dominated when the combined function (s*T) of selection (s) and the length in generations of a particular selective epoch (T) were greater than 10 for both epochs, assuming constant length epochs, or 30 if epoch lengths exhibited an exponential distribution. This suggests that a broad range of selective environments allows for evolution of PV, including short epochs of strong selection combined with long epochs of weak selection (Fig. 2). Critically, the ability to alter mutability via changes in the length of the repeat tract engenders a capacity for localized adaptation, or longer term evolution, of strains to accommodate variations in the prevailing selective environments.

The challenge of a sudden change in environment

Transmission within and between hosts of commensal or pathogenic bacteria will often result in dramatic reductions in bacterial population size.[22] These reductions in population size produce a decrease in genetic diversity of the population precisely at a time when the selective conditions reach their highest stringency, and diversity would be most valuable. Such population contractions may be nonspecific (for example, due to physical processes, such as desiccation or mechanical clearance by mucociliary cells) or may select for a particular genotype (for example, one required for invading host cells). SSR-mediated PV provides a mechanism for rapidly regenerating variation at specific loci. The extent of population contraction is obviously a key element in understanding the impact on population diversity and evolution of phase variable loci. Evidence suggests that a single-cell bottleneck may frequently occur during bacteremic spread of phase variable bacteria, such as *H. influenzae*.[23] Small bottlenecks would select for individuals with very high PV rates, enabling rapid regeneration of the original genetic diversity.

Potential selective forces acting on the alternate ON and OFF states of phase variable loci

Selection for the ON state in most genes will usually be connected with the function of a phase variable determinant, for example, a requirement for adhesion will select for ON expression of adhesins. When multiple loci are involved in a phenotype, such as three biosynthetic enzymes acting on a particular lipopolysaccharide structure,[2] variation of any one can lead to an OFF state. Examples of strong OFF state selectors are adaptive immune responses and bacteriophages. A consideration of how these pressures vary within hosts and after horizontal transfer to a new host is important for understanding how PV may have evolved, and how it mediates ongoing adaptation or evolution during spread of bacteria within different host populations (one key aspect not considered herein is how multiple phase variable genes may contribute to immune evasion during host persistence[24]).

The effect of host immunity on PV during persistence in individuals

The detection of antigen-specific antibodies and the high levels of antigenic variation in surface proteins implies a major impact of adaptive immune responses on commensal and pathogenic bacteria. Losses or reductions in expression of surface determinants facilitate escape from the host immune response. Induction of an effective antibody response should correlate with loss of expression of a phase variable antigen. Although correlations between specific antigens and cognate antibodies have been seen in *Borrelia hermsii*[25] and some parasites, it is less clear if a similar situation is prevalent in phase variable bacterial species. A competitive advantage for OFF variants over ON variants, and strains with high versus low PV rates, can be observed during *in vitro* selection with mAbs specific for phase variable epitopes in meningococci.[26] A key future goal will be to test the hypothesis that persistent infection will lead to induction of antibody responses against phase variable surface antigens and selection for OFF variants.

The effect of host immunity on PV during a spread through a host population

Spread of bacterial species through semi-enclosed populations, during epidemics or in populations with high endemic levels, will generate multiple individuals with full or partial immunity to circulating strains. PV of immunodominant surface antigens may be critical for persistence of bacteria in such populations. The shift from a population with low immunity to one with high immunity will impose a selection for heightened switching rates and multiple phase-variable genes, in a similar fashion, to longitudinal persistence in individual hosts.[9,15] In contrast, a population with a low endemic level of bacteria will require lower switching rates. In this case, the occasional colonization of highly immune individuals, interspersed with long periods in nonimmune individuals, may still provide a strong enough selection for evolution of phase variable loci.

Bacteriophages exert strong selective pressures on surface receptors

Mucosal pathogens may experience a stronger adaptive immune response than do gastrointestinal bacterial species, which tend to be present in higher numbers. These high numbers are advantageous for facilitating host persistence, but provide a significant potential for infection by bacteriophages. As with adaptive immune responses, low levels of phage infection are likely during early stages of host colonization, when the population is small, but will increase as the bacterial population expands and phages become more numerous. This selection may occur during colonization of individual hosts and during spread within a population.[27] Thus, bacterial spread through a population is likely to be followed by waves of infection by distinct bacteriophages. High levels of switching in single genes or PV of multiple phase-variable genes that alter surface receptors for these bacteriophages would facilitate bacterial survival.

Bacteriophages exert selection for phase variable restriction/modification systems

Phase variable restriction–modification systems have evolved that mediate resistance to bacteriophage infection.[28] Multiple systems can inhibit phage spread by generating a diverse population with different combinations of R–M systems in an ON state, whereas single-phase variable R–M systems may generate resistance by a different mechanism, as a methylated phage that replicates in an OFF variant will become unmethylated and thus sensitive to restriction.[29] Oscillation in the numbers of ON/OFF variants and methylated/unmethylated phages may facilitate both bacterial and phage survival. An alternative view is that some phase variable R–M systems have evolved as stochastic regulators (the "phasevarion") of the expression of other genes, which enables adaptation to environmental changes in times of stress.[30]

Summary

Selection for rapid adaptation in environments that repeatedly present distinct challenges, such as host immunity against a surface antigen versus the requirement of that antigen for adhesion to host tissues, has favored the evolution of SSR-mediated PV. The rate and nature of variation in SSRs are affected by the interplay of *cis*-acting factors, such as tract length and repeat unit size, with *trans*-acting factors, such as the replicative machinery and MMR. Secondary selective forces can then act on the variability in these SSRs. Indeed, we have found that the major environmental forces can be modeled as the product of selection strength and length of time before the environment changes. The influence of other forces, such as the dramatic decrease in population size that may accompany arrival in a new individual host, and the effect of distinct combinations of phase variable genes, has yet to be fully assessed. Evidence for the action of these forces has arisen from a combination of experimental measurements of SSR-mutability and theoretical simulations. In conclusion, SSR-mediated PV displays the hallmarks of molecular drivers and population-based selective forces acting together to shape the genome.

Conflicts of interest

The authors declare no conflicts of interest.

References

1. Deitsch, K.W., S.A. Lukehart & J.R. Stringer. 2009. Common strategies for antigenic variation by bacterial, fungal and protozoan pathogens. *Nat. Rev. Microbiol.* **7**: 493–503.
2. Moxon, E.R., C.D. Bayliss & D.W. Hood. 2007. Bacterial contingency loci: the role of simple sequence DNA repeats in bacterial adaptation. *Ann. Rev. Genet.* **40**: 307–333.
3. van der Woude, M.W. & A.J. Baumler. 2004. Phase and antigenic variation in bacteria. *Clin. Microbiol. Rev.* **17**: 581–611.
4. Streisinger, G. *et al.* 1966. Frameshift mutations and the genetic code. *Cold Spring Harb. Symp. Quant. Biol.* **31**: 77–84.
5. Levinson, G. & G. Gutman. 1987. Slipped-strand mispairing: a major mechanism for DNA sequence evolution. *Mol. Biol. Evol.* **4**: 203–221.
6. Bayliss, C.D. *et al.* 2012. Phase variable genes of *Campylobacter jejuni* exhibit high mutation rates and specific mutational patterns but mutability is not the major determinant of population structure during host colonization. *Nuc. Acids Res.* **33**: 504–520.
7. van Ham, S.M. *et al.* 1993. Phase variation of *H. influenzae* fimbriae: transcriptional control of two divergent genes through a variable combined promoter region. *Cell* **73**: 1187–1196.
8. Barry, J.D., J.P.J. Hall & L. Plenderleith. 2012. Genome hyperevolution and the success of a parasite. *Ann. N.Y. Acad. Sci.* **1267**: 11–17. This volume.
9. Bayliss, C.D. 2009. Determinants of phase variation rate and the fitness implications of differing rates for bacterial pathogens and commensals. *FEMS Microbiol. Rev.* **33**: 504–520.
10. Bayliss, C.D., T. van de Ven & E.R. Moxon. 2002. Mutations in *polI* but not *mutSLH* destabilize *Haemophilus influenzae* tetranucleotide repeats. *EMBO J.* **21**: 1465–1476.
11. Kroutil, L.C. *et al.* 1996. Exonucleolytic proofreading during replication of repetitive DNA. *Biochemistry* **35**: 1046–1053.
12. Tran, H.T. *et al.* 1997. Hypermutability of homonucleotide runs in mismatch repair and DNA polymerase proofreading yeast mutants. *Mol. Cell. Biol.* **17**: 2859–2865.
13. Morel, P. *et al.* 1998. The role of SOS and flap processing in microsatellite instability in *Escherichia coli*. *Proc. Natl. Acad. Sci. USA* **95**: 10003–10008.
14. Gawel, D. *et al.* 2002. Asymmetry of frameshift mutagenesis during leading and lagging-strand replication in *Escherichia coli*. *Mutat. Res.* **501**: 129–136.
15. Richardson, A.R. *et al.* 2002. Mutator clones of *Neisseria meningitidis* in epidemic serogroup A disease. *Proc. Natl. Acad. Sci. USA* **99**: 6103–6107.
16. Martin, P. *et al.* 2004. Involvement of genes of genome maintenance in the regulation of phase variation frequencies in *Neisseria meningitidis*. *Microbiology* **150**: 3001–3012.
17. De Bolle, X. *et al.* 2000. The length of a tetranucleotide repeat tract in *Haemophilus influenzae* determines the phase variation rate of a gene with homology to type III DNA methyltransferases. *Mol. Microbiol.* **35**: 211–222.
18. Lin, W.H. & E. Kussell. 2011. Evolutionary pressures on simple sequence repeats in prokaryotic coding regions. *Nucleic Acids Res.* **40**: 2399–2413.
19. Kumar, P. & H.A. Nagarajaram. 2012. A study on mutational dynamics of simple sequence repeats in relation to mismatch repair system in prokaryotic genomes. *J. Mol. Evol.* **74**: 127–139.
20. King, D. 2012. Indirect selection of implicit mutation protocols. *Ann. N.Y. Acad. Sci.* **1267**: 45–52. This volume.
21. Moxon, E.R. *et al.* 1994. Adaptive evolution of highly mutable loci in pathogenic bacteria. *Curr. Biol.* **4**: 24–33.
22. Meynell, G.G. & B.A. Stocker. 1957. Some hypotheses on the aetiology of fatal infections in partially resistant hosts and their application to mice challenged with *Salmonella paratyphi-B* or *Salmonella typhimurium* by intraperitoneal injection. *J. Gen. Microbiol.* **16**: 38–58.
23. Moxon, E.R. & P.A. Murphy. 1978. *Haemophilus influenzae* bacteremia and meningitis resulting from survival of a single organism. *Proc. Natl. Acad. Sci. USA* **75**: 1534–1536.
24. Bayliss, C.D., D. Field & E.R. Moxon. 2001. The simple sequence contingency loci of *Haemophilus influenzae* and *Neisseria meningitidis*. *J. Clin. Invest.* **107**: 657–662.
25. Barbour, A.G. *et al.* 2006. Pathogen escape from host immunity by a genome program for antigenic variation. *Proc. Natl. Acad. Sci. USA* **103**: 18290–18295.
26. Bayliss, C.D. *et al.* 2008. *Neisseria meningitidis* escape from the bactericidal activity of a monoclonal antibody is mediated by phase variation of *lgtG* and enhanced by a mutator phenotype. *Infect Immun.* **76**: 5038–5048.
27. Connerton, P.L., A.R. Timms & I.F. Connerton. 2011. Campylobacter bacteriophages and bacteriophage therapy. *J. Appl. Microbiol.* **111**: 255–265.
28. Zaleski, P., M. Wojciechowski & A. Piekarowicz. 2005. The role of Dam methylation in phase variation of *Haemophilus influenzae* genes involved in defence against phage infection. *Microbiology* **151**: 3361–3369.
29. Bayliss, C.D., M.J. Callaghan & E.R. Moxon. 2006. High allelic diversity in the methyltransferase gene of a phase variable type III restriction-modification system has implications for the fitness of *Haemophilus influenzae*. *Nucleic Acids Res.* **34**: 4046–4059.
30. Srikhanta, Y.N., K.L. Fox & M.P. Jennings. 2010. The phasevarion: phase variation of type III DNA methyltransferases controls coordinated switching in multiple genes. *Nat. Rev. Microbiol.* **8**: 196–206.

Ann. N.Y. Acad. Sci. ISSN 0077-8923

ANNALS OF THE NEW YORK ACADEMY OF SCIENCES
Issue: *Effects of Genome Structure and Sequence on Variation and Evolution*

Indirect selection of implicit mutation protocols

David G. King

Department of Anatomy and Department of Zoology, Southern Illinois University Carbondale, Carbondale, Illinois

Address for correspondence: David G. King, Department of Anatomy 6512, SIUC, Carbondale, IL 62901. dgking@siu.edu

A hypothesis that mutability evolves to facilitate evolutionary adaptation is dismissed by many biologists. Their skepticism is based on a theoretical expectation that natural selection must minimize mutation rates. That view, in turn, is historically grounded in an intuitive presumption that "the vast majority of mutations are harmful." But such skepticism is surely misplaced. Several highly mutagenic genomic patterns, including simple sequence repeats, and transposable elements, are integrated into an unexpectedly large proportion of functional genetic loci. Because alleles arising within such patterns can retain an intrinsic propensity toward a particular style of mutation, natural selection that favors any such allele can indirectly favor the site's mutability as well. By exploiting patterns that have produced beneficial alleles in the past, indirect selection can encourage mutation within constraints that reduce the probability of deleterious effect, thereby shaping implicit "mutation protocols" that effectively promote evolvability.

Keywords: mutation rate; variation; evolvability; indirect selection; protocols; simple sequence repeats

[O]ne point, which has greatly troubled me; . . . what the devil determines each particular variation? What makes a tuft of feathers come on a cock's head, or moss on a moss-rose?[1]

Charles Darwin

Charles Darwin's deep concern with such questions led him to predict, "A grand and almost untrodden field of inquiry will be opened, on the causes and laws of variation."[2] He surely would have appreciated the theme of this volume,[a] that the causes and laws of variation include some fascinating features of genome organization that actually encourage certain styles of mutation.

Historical emphasis on mutations as haphazard accidents

From the beginning of modern genetics, mutations have been regarded as errors in the transmission of hereditary information. And, as Calvin Bridges (one of the pioneers of *Drosophila* genetics) explained over ninety years ago, mistakes are unlikely to be advantageous.

Any organism as it now exists must be regarded as a very complex physicochemical machine with delicate adjustments of part to part. Any haphazard change made in this mechanism would almost certainly result in a decrease of efficiency. . . . Only an extremely small proportion of mutations may be expected to improve a part or the interrelation of parts in such a way that the fitness of the whole organism for its available environments is increased.[3]

Despite its entirely intuitive origin, this conviction has been promulgated with remarkable consistency throughout the past century [emphasis added]:

- 1909, Francis Galton (Charles Darwin's cousin): "[T]he *vast majority* of mutations end up reducing the number of offspring."[4]
- 1930, J.B.S. Haldane (who helped establish the modern synthesis): "The *vast majority* of mutations are harmful."[5]
- 1937, Alfred Sturtevant (one of the pioneers of *Drosophila* genetics): "The *vast majority* of mutations are unfavorable."[6]
- 1989, John Maynard Smith (one of the founders of evolutionary game theory): "It is *common sense* that most mutations that alter fitness at all will lower it."[7]

---

[a]Effects of Genome Structure and Sequence on the Generation of Variation and Evolution. 2012. *Annals of the New York Academy of Sciences*. Volume 1267.

doi: 10.1111/j.1749-6632.2012.06615.x
Ann. N.Y. Acad. Sci. 1267 (2012) 45–52 © 2012 New York Academy of Sciences.

- 2007, C. Baer *et al.* (reviewing recent analyses of mutation rate evolution): "[T]he *vast majority* of mutations with observable effects are deleterious."[8]

Yet such assurance is founded on surprisingly little data, primarily from experiments with radiation and chemical mutagens. But although artificially induced mutations are indeed predominantly deleterious, for mutations arising spontaneously under natural conditions the ratio of benefit to harm has never been realistically assessed.[9] Nevertheless, a presumed preponderance of deleterious mutations has informed most analyses of mutation rate evolution. As first argued by Alfred Sturtevant:

> [F]or every favorable mutation, the preservation of which will tend to increase the number of genes in the population that raises the mutation rate, there are hundreds of unfavorable mutations that will tend to lower it. Further, the unfavorable mutations are mostly highly unfavorable, and will be more effective in influencing the rate than will the relatively slight improvements that can be attributed to the rare favorable mutations. This raises the question—why does the mutation rate not become reduced to zero? No answer seems possible at present, other than the surmise that the nature of genes does not permit such a reduction. *In short, mutations are accidents, and accidents will happen.*[6] [emphasis added]

George Williams ("widely regarded...as one of the most influential and incisive evolutionary theorists of the 20th century"[10]) did not mince words when he reiterated Sturtevant's argument in 1966:

> The fittest possible degree of stability is *absolute* stability. In other words, natural selection of mutation rates has *only one possible direction*, that of reducing the frequency of mutation *to zero*. . . .Evolution has probably reduced mutation rates to far below species optima, as the result of *unrelenting selection* for zero mutation rate in every population. . . .So evolution takes place, not so much because of natural selection, but to a large degree in spite of it.[11] [emphasis added]

Although challenged repeatedly in recent decades,[12–16] this conclusion remains influential. One recent review flatly declared, "the cost of fidelity is the generally accepted explanation for nonzero mutation rates in multicellular eukaryotes."[8]

Protocols for mutation

The contrary hypothesis, that facilitated variation is a true genomic function, also has deep historical roots. Darwin himself observed that "[s]ome authors believe it to be as much the function of the reproductive system to produce individual differences, or very slight deviations of structure, as to make the child like its parents."[17] In her eloquent preface to this volume's predecessor, Lynn Caporale took a more radical view based on consideration that every evolving lineage faces an ever-changing selective landscape:

> Far from clumsy stumblers into random point mutations, genomes have evolved mechanisms that facilitate their own evolution. These mechanisms . . .diversify a genome and increase the probability that its descendants will survive.[12]

More succinctly, "Life has evolved to evolve."[15] Indeed, a propensity to mutate within appropriate constraints could be a fundamental attribute of most genomes.

From the orthodox perspective rehearsed above, such assertions appear naive. But the keystone assumption of orthodoxy, that the vast majority of mutations are deleterious, ignores the multiplicity of distinct mutational mechanisms whose fitness-effect probabilities vary over differing genomic domains.[18] What if advantageous constraints could be imposed on certain mechanisms, to shift the overall impact of mutation toward a more favorable balance between benefit and harm? (The term "constraints" commonly implies limitations rather than creative opportunities, but constraints that suppress a sufficiently large number of unpromising possibilities can thereby increase the accessibility of more favorable alternatives.) The result would be implicit "mutation protocols," genomic patterns that permit localized increases in variability. Mutations that abide by protocol would not be random with respect to mechanism or genomic location. But mutations need not be accidental or haphazard to remain random in the sense required by classical Darwinian theory, that specific variants not be directed toward particular adaptive ends.

"Protocols" in this sense, as introduced by Csete and Doyle to address the robustness and fragility that characterize complex living systems, are "rules designed to manage relationships and processes smoothly and effectively:"[19]

> Thinking in terms of protocols, in addition to genes, organisms, and populations, as foci of natural selection, may be a useful abstraction for understanding the evolution of complexity [cf. Ref. 13]. Good protocols allow new functions to be built from existing components and allow new components to be added or to evolve from existing ones, powerfully enhancing both engineering and evolutionary "tinkering."[19]

Examples of mutation protocols with undoubted adaptive value include antigenic variation in parasitic microorganisms[20] and hypervariability of vertebrate antibody genes.[21] But the "useful abstraction" of a mutation protocol is perhaps most clearly illustrated by a much simpler example, a "tuning knob" protocol implemented by simple sequence repeats.

The tuning knob protocol

Simple sequence repeats (SSRs), in which a simple DNA motif is repeated several times in tandem (e.g., CACACACA, CAGCAGCAG, or CAATCAAT-CAAT), are a common feature in both prokaryotic and eukaryotic genomes. SSRs yield quantitative genetic variation that is both abundant and relatively safe. The following features are all well documented:[22]

- *SSRs implement a specific style of mutation.* They increase or decrease the number of repeating motifs, commonly by a single motif unit, without otherwise altering the repetitive pattern.
- *SSRs have high, site-specific mutation rates.* These are based on mechanisms that include replication slippage and unequal recombination. Mutation frequency at any particular site can be several orders of magnitude higher than the genome-wide average for base-pair substitution and depends on local sequence features such as motif length, number of repeats, and purity of repetition.
- *SSR mutations yield small, incremental phenotypic effects.* Although repeat variation has often been regarded as effectively neutral, numerous studies demonstrate that repeat-number variants can be causally associated with quantitative variation in gene function. (Extreme ex-

amples with pathological effects are also well known.)
- *SSR mutations are readily reversible.* Any stepwise change in the number of repeats can be undone by a step in the opposite direction.
- *SSRs are modular.* Each instance is intrinsically linked to a particular genomic site, has its own genetic effects and parameters for variation, and can evolve independently from any other repeat site.

These features, which together permit efficient incremental adjustment of gene function, have led to metaphorical characterization of SSRs as "evolutionary tuning knobs."[23]

Adjustment by tuning knob represents a huge improvement over haphazard accident. If a violin string is out of tune, a small random twist of its tuning knob has a much higher probability of improving the tone (about fifty percent) than would a sudden, accidental impact. A vast majority of such small adjustments should not be substantially deleterious, and the same knob can be repeatedly adjusted without risk to other structures. Of course incremental adjustability does not require a literal knob, just some physical configuration that can set a specific parameter value while also facilitating small changes of that value. For any SSR, the "setting" is a particular repeat-number allele, while "adjustment" is accomplished by reversible repeat-number mutations. As with a literal knob, no matter how often the repeat number changes, the SSR remains fundamentally adjustable.

Far from being a rare feature of special genes, the tuning knob protocol is ubiquitous. A typical eukaryotic genome contains hundreds of thousands of SSRs. They occur in a high proportion of genes and can be located in exons or introns, in transcribed or untranscribed regions, and in upstream or downstream regulatory regions.[22] Extended genes can include several distinct SSR sites within their regulatory and/or coding domains, so that practically any aspect of genetic function can be adjusted by associated SSR variation. And the tuning knob protocol can be implemented by a wide assortment of repeats based on many different sequence motifs. (Depending on motif and context, SSRs can also behave as reversible switches that turn gene function on or off, as in many bacterial contingency genes.[24])

Of course, not all random adjustments will be advantageous, so each site with an active protocol should impose some selective burden. But only mutations in the wrong direction will be deleterious, and since tuning knob mutations yield mostly small, quantitative phenotypic effects,[22] most deleterious mutations will be only slightly disadvantageous. Furthermore, as recognized by R. A. Fisher over eighty years ago, the proportion of "mutations of small effect" that are deleterious will be closer to fifty percent rather than a "vast majority," at least under unexceptional circumstances of suboptimal adaptation during environmental change.[25] Thus, the associated fitness cost can remain low and relatively constant even while a steady supply of new variants ensures reliable response to shifting selection pressures, including rapid replacement of potentially beneficial alleles that are lost to genetic drift. By contrast, the cost of reproducing without abundant selectable variation (i.e., with new variants appearing only as accidental, haphazard errors) can increase very quickly during extensive shifts of an adaptive peak, if a population must endure declining fitness while awaiting the appearance of each improbably beneficial "mistake."

The SSR mutation protocol bridges the gap between epigenetic responses that are rapid but not reliably inherited across generations and those arising from conventional single nucleotide substitutions that occur at a much slower rate but have much greater stability.[22] Tuning knob effects have been implicated as the basis for ongoing adaptation in several natural populations. For example, the number of threonine-glycine and serine-glycine dipeptide repeats in the *Drosophila* PER protein, encoded by a polymorphic hexanucleotide SSR in exon 5 of the *period* gene, influences sensitivity of circadian rhythm to temperature fluctuations.[26] Geographic and microgeographic clines in the frequencies of repeat-number alleles of this gene have been interpreted as evidence for adaptively fine-tuning this gene to local climate.[27] A similar association between geography and repeat variation has been reported for an avian clock gene.[28] In human populations, locally elevated mutation rates at a short thymidine repeat in the heart disease gene *MMP3* yield variation that experiences strong positive selection.[29] Experimental studies with yeast have directly demonstrated functional variability arising from intragenic tandem repeats[30] as well as advantageous transcriptional evolvability based on enrichment of SSRs in gene promotors.[31]

Indirect selection

To the extent that SSRs (as well as other prolific sources of variation) have been recognized as important contributors to adaptive evolution, this role has been generally perceived as fortuitous. But if a mutation protocol can increase the speed or effectiveness of adaptation—that is, if systematically constrained variation offers a more efficient route to discovery of beneficial variants—then perhaps, in contrast to traditional orthodoxy, locally elevated mutation rates can also be indirectly favored by selection.[32–35]

To see why selection of mutation protocols must necessarily be indirect, we revisit our metaphor. A violinist's audience listens not for the presence of tuning knobs but for music that is played in tune. Yet applause for a tuneful performance must also, indirectly, be applause for the tunable violin. Similarly, the agencies of natural selection, conceived as acting on individual organisms, cannot directly discriminate between better and worse mutation protocols. Selection sees only how well each particular phenotype is suited for its immediate environment. Thus, only particular genetic parameters (i.e., individual repeat-number alleles) are selected directly, not any associated potential for adjustability. Nevertheless, whenever selection favors a particular SSR allele, that selection must also, indirectly, favor the mutation protocol that is implicit in motif repetition.

Just like direct selection, indirect selection acts in every generation upon every allele; it is no less effective because it is indirect. Furthermore, while a direct selective sweep based on a single beneficial allele can indirectly establish an incipient SSR throughout a population, the incipient protocol can be suppressed only very gradually as direct selection eliminates deleterious mutations as they arise, one allele at a time. Thus an effective mutation protocol can be favored over the long term even when most associated mutations are not advantageous (i.e., even when direct selection would seem to oppose its mutagenicity), as long as fitness-enhancing alleles arise often enough to override the associated direct cost of deleterious mutations.

The processes that create, maintain, and eliminate an SSR are closely related. Single base-pair substitutions, insertions, or deletions can convert a nonrepetitive sequence into a short SSR or introduce interruptions into a preexisting SSR; interruptions can be eliminated by subsequent slippage mutations that extend and shorten a repeat tract. The balance over time among all such mutations, together with the effects of selection on particular alleles, will determine the evolutionary duration ("life cycle") of the SSR.[36] Maximally stable alleles should outcompete tuning knob alleles only when variation associated with the protocol is consistently disadvantageous for very many generations, in which case orthodox theory correctly predicts that selection should push mutation rates downward until costs of replication fidelity balance lost reproductive capacity from deleterious mutations. But the only circumstance necessary for the tuning knob protocol to prevail is a selective environment in which constrained variation is occasionally advantageous.

Indirect selective advantage has been documented for the SSR-based mutability of contingency loci in pathogenic prokaryotes.[24,37] But confidence in orthodoxy has hindered appreciation for this example's broader significance:

> Interestingly, unstable sequence features such as tandem repeats tend to be found disproportionately near and within [bacterial contingency loci], where they enhance variability and remain linked to new beneficial mutations. Contingency loci provide our best example of evolvability-as-adaptation, but their relevance to the general question of whether evolvability-as-adaptation is a major missing component of evolutionary theory seems rather limited. . . . It is indeed attractive to suppose that the most important evolutionary feature of organisms—their very capacity to evolve and adapt—is itself an adaptation, but this is probably only true in highly restricted circumstances.[38]

Circumstances in which SSR mutability is advantageous can seem "highly restricted" only if one ignores the ubiquity of SSRs throughout most eukaryotic genomes, their functional integration into a high proportion of genes, and the conservation of some variable SSR sites across long expanses of evolutionary time. Taken together, these all suggest that the SSR mutation protocol, shaped by indirect selection, could indeed play a major role in organisms' "very capacity to evolve and adapt."[22,23,31–35]

Additional mutation protocols

While the SSR-based tuning knob protocol supports adaptive adjustment of preexisting genetic functions, novel functions can emerge through opportunities provided by a "copy-and-paste" protocol implemented by transposable elements (TEs). TE activity can even be conveniently integrated with the tuning knob protocol. Some TEs use SSRs as sites for insertion (i.e., as "landing pads"[39,40]), while many TE insertions create new repeats at either end. Thus, if the action of a transposable element creates a new function, that function can come with ready-made tuning knobs for its adjustment. TEs are often disparaged as selfish or parasitic because their activity can disrupt gene function. Orthodoxy presumes that selection at the organism level must thus act in only one direction, toward suppression of TE mobility. But since TEs play an acknowledged role in genome evolution, creating permissive and possibly necessary conditions for adaptive innovation and diversification,[41–44] TEs might instead become "domesticated"[41] by indirect selection so potential harm is reduced while evolutionary utility is retained. An apparently selfish tendency toward dispersal throughout a genome can prolong the time available for such domestication, since a widely distributed TE stands a higher probability of producing at least one advantageous mutation, and hence becoming more firmly established, before its mutagenic potential can be suppressed.

The above examples of tuning knob and copy-and-paste protocols only hint at the potential consequences of indirect selection. Among the most familiar processes for generating systematic variation are those of sexual reproduction. Yet these have been explicitly excluded from textbook definitions of mutation (e.g., [emphasis added] "*mutation*: an *error* in replication of a nucleotide sequence, or any other alteration of the genome that is *not* manifested as reciprocal recombination"[45]) precisely because the vast majority of recombination products are *not* deleterious. Instead of being categorized as something distinct from mutation, meiotic recombination could be more productively conceptualized as perhaps the most sophisticated and highly organized of mutation protocols. If evolutionary processes can create and maintain sources of variation as elaborate as those of sexual reproduction, despite their steep cost, then additional protocols for generating potentially advantageous variation can surely be expected. Intriguing possibilities include protocols that permit hypermutation at particular sites, such as genes mediating interspecies interaction;[46,47] protocols (in addition to retrotransposition) for copy number variation;[48] and protocols that organize recombination hotspots, coldspots, and inversions.[49–51] Recent demonstration of non-random variation of local mutation rates in *E. coli* (e.g., lower rates in highly expressed genes) suggests localized mutation-rate optimization based on mechanisms yet to be elucidated.[52] Eukaryotic genomes are unlikely to be any less sophisticated.

Counterarguments

Defenders of evolutionary orthodoxy have maintained that the hypothesis of advantageous mutation protocols is unnecessary:

> One might call this the evolvability-as-adaptation hypothesis. This hypothesis is intuitively appealing, but it has its own set of problems. For one thing, it assumes that variability—the capacity to generate new variation—is generally limiting to population persistence and success in nature. This assumption lacks empirical support. Most natural populations have large amounts of standing genetic variation and do not necessarily depend on *de novo* variation in order to adapt to environmental change. Directly observed rates of short-term evolution in natural populations often far exceed those inferred from the fossil record, and this too implies that there is ample capacity for adaptation in response to selection in most populations.[38]

But this argument fails by simply assuming the very phenomenon that mutation protocols help to explain: the reliable abundance of adaptively appropriate variation. A hypothesis that indirect selection might favor mutation protocols does not require that variability be generally limiting in most populations, only that the evident adequacy of standing variation might call for explanation beyond haphazard accident from imperfect replication. The concept of mutation protocols implies that serviceable variation abounds *because* effective mutation protocols have become thoroughly incorporated into most genomes.

A related counterargument holds that "evolvability-as-adaptation must be the con-

sequence of selection among populations rather than selection among individuals."[38] The widely accepted weakness of population-level selection (relative to individual selection, which by orthodoxy opposes mutability) then impugns the plausibility of selection for evolvability. But this argument depends on conflating evolvability *per se* with the underlying molecular and developmental protocols that confer evolvability.[35] Evolvability is indeed (by definition) a property of populations. But indirect selection does not require competition among populations based on differences in evolvability. Advantageous mutation protocols are selected indirectly (and powerfully) at the level of individual alleles and organisms, whenever direct selection favors a mutation that arises according to protocol.

The term "lineage selection" does offer a convenient shorthand for appreciating the relationship between indirect, organism-level selection of mutation protocols and the emergence of evolvability.[44] By increasing the efficiency of adaptation, an effective mutation protocol would increase a lineage's chance of enduring over time (i.e., the lineage analog of individual survival), while also boosting its prospects for adaptive radiation (i.e., the lineage analog of reproduction). Nevertheless, although lineage selection could plausibly reinforce the effects of direct or indirect selection acting at the conventional, organismal level,[53] actual competition among separate lineages is neither necessary nor sufficient to explain the establishment of mutation protocols. Just like any other trait that potentially affects fitness, each embodiment of a protocol must stand or fall based on how well its alleles support each individual's contribution to the next generation. Increasingly efficient evolvability emerges as an inevitable consequence when mechanisms of mutation become more advantageously constrained by effective mutation protocols.[33–35]

Implications for comparative genomics

The hypothesis presented in this essay—that indirect selection for the constrained variation of implicit mutation protocols has promoted faster and safer exploration of genome sequence space than would be possible by haphazard accident alone—carries an interesting corollary. Those sequence patterns that are most important for adaptive adjustment and evolutionary innovation might be among those that are most mutable and hence least conserved over evolutionary time. Although "conservation equals function" is a reasonable expectation in certain contexts, most biologists also understand that many species-specific traits can contribute to fitness without being conserved at higher taxonomic levels. Comparative genomics should not focus exclusively on sequence conservation, precisely because unstable, evolutionarily labile sites could be especially relevant for understanding ongoing adaptation.[37,47,54] An exponential decline in the proportion of SSRs that are conserved over increasing phylogenetic distance has been interpreted as evidence for neutral evolution of such sites in the face of mutation pressure and drift.[55] But such a decline is also consistent with temporal variation in the circumstances for indirect selection at each particular SSR, for example, during periods of niche divergence or environmental change. Even so, some repeat sites have been deeply conserved.[55] One especially intriguing set of dinucleotide repeat sites is shared between human and opossom, even while mutations have churned the sites sufficiently to alter their motifs and thus obscure their homology in other species;[56] curiously, most of these sites occur within genes having neurodevelopmental roles.[57] Perhaps the simplest explanation for such persistently mutable sites could be their durable utility as a substrate for efficient adaptive adjustment of behavior.

Conclusion

A presumption that mutations are haphazard accidents, whose low probability of beneficial effect assures that selection must minimize mutation rates, is challenged by a significant body of data concerning several prolific sources of genetic change, notably simple sequence repeats, transposable elements, and sexual recombination. Just as life itself depends on ancient and strongly conserved metabolic protocols, such as those for using ATP and synthesizing proteins, so also could efficient adaptation in diverse and ever-changing environments depend on implicit "mutation protocols" that have become established throughout most genomes.[58] This hypothesis has broad implications, both for enduring controversies in evolutionary theory and for current computational analyses in comparative genomics. Studies in anatomy, physiology, and biochemistry have long relied on the assumption that organisms are adaptively "designed" by natural selection.

We should not be surprised if the creative power of indirect selection parallels that of direct natural selection. Once released from the obsolete and misleading presumption that mutability cannot be advantageous, genetics should also advance by reverse engineering[19] those molecular mechanisms for mutation that not only serve immediate adaptation but also promote continuing evolvability.

Conflicts of interest

The author declares no conflicts of interest.

References

1. Darwin, C. 1892. *Charles Darwin, His Life Told in an Autobiographical Chapter and in a Selected Series of his Published Letters*, Dover ed. Francis Darwin. D. Appleton and Co, Ed.: 227. 1958. Dover Publications Inc. New York.
2. Darwin, C. 1859. *On the Origin of Species by Means of Natural Selection.* John Murray. London. (Facsimile edition, p. 486. 1964. Harvard University Press. Cambridge, Massachusetts.)
3. Bridges, C.B. 1919. Specific modifiers of eosin eye color in *Drosophila melanogaster. J. Exp. Zool.* **28:** 337–384.
4. Galton, F. 1909. *Essays in Eugenics.* Eugenics Education Society. London.
5. Haldane, J.B.S. 1930. *Possible Worlds and Other Essays*: 36. Chatto & Windus. London.
6. Sturtevant, A.H. 1937. Essays on evolution: I. On the effects of selection on mutation rate. *Q. Rev. Biol.* **12:** 464–467.
7. Maynard Smith, J. 1989. *Evolutionary Genetics*: 55. Oxford University Press. Oxford.
8. Baer, C.F., M.M. Miyamoto & D.R. Denver. 2007. Mutation rate variation in multicellular eukaryotes: causes and consequences. *Nat. Rev. Genet.* **8:** 619–631.
9. Shaw, R.G., F.H. Shaw & C. Geyer. 2003. What fraction of mutations reduces fitness? *Evolution* **57:** 686–689.
10. Futuyma, D., as quoted by N. Wade. 2010. *The New York Times* <http://www.nytimes.com/2010/09/14/science/14williams.html>.
11. Williams, G.C. 1966. *Adaptation and Natural Selection*: 138–139. Princeton University Press. Princeton.
12. Caporale, L.H. 1999. Molecular strategies in biological evolution. *Ann. N.Y. Acad. Sci.* **870:** xi–xii.
13. Kirschner, M. & J. Gerhart. 1998. Evolvability. *Proc. Natl. Acad. Sci. USA* **95:** 8420–8427.
14. Caporale, L.H. 2003. Natural Selection and the emergence of a mutation phenotype: an update of the evolutionary synthesis considering mechanisms that affect genome variation. *Annu. Rev. Microbiol.* **57:** 467–485.
15. Earl, D.J. & M.W. Deem. 2004. Evolvability is a selectable trait. *Proc. Nat. Acad. Sci. USA* **101:** 11531–11536.
16. Gerhart, J. & M. Kirschner. 2007. The theory of facilitated variation. *Proc. Natl. Acad. Sci. USA* **104:** 8582–8589.
17. Darwin, C. 1859. *On the Origin of Species by Means of Natural Selection.* John Murray. London. (Facsimile edition, p. 131. 1964. Harvard University Press. Cambridge, Massachusetts.)

18. King, D.G. & Y. Kashi. 2007. Mutation rate variation in eukaryotes: evolutionary implications of site-specific mechanisms. *Nature Rev. Genet.* **8** doi: 10.1038/nrg2158-c1. http://www.nature.com/nrg/journal/v8/n11/full/nrg2158-c1.html.

19. Csete, M.E. & J.C. Doyle. 2002. Reverse engineering of biological complexity. *Science* **295:** 1664–1669.

20. Barry, D.J. 2006. Implicit information in eukaryotic pathogens as the basis of antigenic variation. In: *The Implicit Genome.* Lynn H. Caporale, Ed.: 91–106. Oxford University Press. New York.

21. Beale, R. & D. Iber. 2006. Somatic evolution of antibody genes. In: *The Implicit Genome.* Lynn H. Caporale, Ed.: 177–190. Oxford University Press. New York.

22. Gemayel, R., M.D. Vinces, M. Legendre & K.J. Verstrepen. 2010. Variable tandem repeats accelerate evolution of coding and regulatory sequences. *Annu. Rev. Genet.* **44:** 445–477.

23. King, D.G., Y. Kashi & M. Soller. 1997. Evolutionary tuning knobs. *Endeavour* **21:** 36–40.

24. Bayliss, C.D. & E.R. Moxon. 2006. Repeats and variation in pathogen selection. In: *The Implicit Genome.* Lynn H. Caporale, Ed.: 54–76. Oxford University Press. New York.

25. Fisher, R.A. 1930. *The Genetical Theory of Natural Selection.* Oxford University Press. Oxford.

26. Sawyer, L.A., J. M. Hennessy, A.A. Peixoto, *et al.* 1997. Natural variation in a *Drosophila* clock gene and temperature compensation. *Science* **278:** 2117–2120.

27. Kyriacou, C.P., A.A. Peixoto, F. Sandrelli, *et al.* 2008. Clines in clock genes: fine-tuning circadian rhythms to the environment. *Trends Genet.* **24:** 124–132.

28. Johnsen, A., A.E. Fidler, S. Kuhn, *et al.* 2007. Avian *Clock* gene polymorphism: evidence for a latitudinal cline in allele frequencies. *Mol. Ecol.* **16:** 4867–4880.

29. Rockman, M.V., M.W. Hahn, N. Soranzo, *et al.* 2004. Positive selection on *MMP3* regulation has shaped heart disease risk. *Current Biol.* **14:** 1531–1539.

30. Verstrepen, K.J., A. Jansen, F. Lewitter & G.R. Fink. 2005. Intragenic tandem repeats generate functional variability. *Nature Genet.* **37:** 986–990.

31. Vinces, M.D., M. Legendre, M. Caldara, *et al.* 2009. Unstable tandem repeats in promoters confer transcriptional evolvability. *Science* **324:** 1213–1216.

32. King, D.G. & M. Soller. 1999. Variation and fidelity: The evolution of simple sequence repeats as functional elements in adjustable genes. In: *Evolutionary Theory and Processes: Modern Perspectives.* S.P. Wasser, Ed.: 65–82. Kluwer Academic Publishers. Dordrecht, The Netherlands.

33. King, D.G., E.N. Trifonov & Y. Kashi. 2006. Tuning knobs in the genome: evolution of simple sequence repeats by indirect selection. In: *The Implicit Genome.* Lynn H. Caporale, Ed.: 77–90. Oxford University Press. New York.

34. Kashi, Y. & D.G. King. 2006. Has simple sequence repeat mutability been selected to facilitate evolution? *Isr. J. Ecol. Evol.* **52:** 331–342.

35. King, D.G. & Y. Kashi. 2007. Indirect selection for mutability. *Heredity* **99:** 123–124.

36. Buschiazzo, E. & N.J. Gemmel. 2006. The rise, fall and renaissance of microsatellites in eukaryotic genomes. *BioEssays* **28:** 1040–1050.

37. Bayliss, C.D. & M.E. Palmer. 2012. Evolution of simple sequence repeat-mediated phase variation in bacterial genomes. *Ann. N.Y. Acad. Sci.* **1267:** 39–44. This volume.

38. Sniegowski, P.D. & H.A. Murphy. 2006. Evolvability. *Current Biol.* **16:** R831-R834.

39. Nadir, E., H. Margalit, T. Gallily & S.A. Ben-Sasson. 1996. Microsatellite spreading in the human genome: evolutionary mechanisms and structural implications. *Proc. Natl. Acad. Sci. USA* **93:** 6470–6475.

40. Ramsay, L., M. Macaulay, L. Cardle, *et al.* 1999. Intimate association of microsatellite repeats with retrotransposons and other dispersed repetitive elements in barley. *Plant J.* **17:** 415–425.

41. Volff, J.-N. 2006. Turning junk into gold: domestication of transposable elements and the creation of new genes in eukaryotes. *BioEssays* **28:** 913–922.

42. Jurka, J., V.V. Kapitonov, O. Kohany & M.V. Jurka. 2007. Repetitive sequences in complex genomes: structure and evolution. *Annu. Rev. Genomics Hum. Genet.* **8:** 241–259.

43. Jurka, J. 2007. Conserved eukaryotic transposable elements and the evolution of gene regulation. *Cell. Mol. Life Sci.* **65:** 201–204.

44. Oliver, K.R. & W.K. Greene. 2009. Transposable elements: powerful facilitators of evolution. *BioEssays* **31:** 703–714.

45. Futuyma, D.J. 1998. *Evolutionary Biology,* 3rd ed.: 769. Sinauer Associates. Sunderland, Massachusetts.

46. Olivera, B.M. 2006. Conus peptides: biodiversity-based discovery and exogenomics. *J. Biol. Chem.* **281:** 31173–31177.

47. Olivera, B.M., M. Watkins, P. Bandyopadhyay, *et al.* 2012. Adaptive radiation of venomous marine snail lineages and the accelerated evolution of venom peptide genes. *Ann. N.Y. Acad. Sci.* **1267:** 61–70. This volume.

48. Schrider, D.R. & M.W. Hahn. 2010. Gene copy-number polymorphism in nature. *Proc. R. Sci. B.* **277:** 3213–3221.

49. Noor, M.A.F., K.L. Grams, L.A. Bertucci & J. Reiland. 2001. Chromosomal inversions and the reproductive isolation of species. *Proc. Natl. Acad. Sci. USA* **98:** 12084–12088.

50. Butlin, R.K. 2005. Recombination and speciation. *Molec. Ecol.* **14:** 2621–2635.

51. Smukowski, C.S. & M.A.F. Noor. 2011. Recombination rate variation in closely related species. *Heredity* **107:** 496–508.

52. Martincorena, I., A.S.N. Seshasayee & N.M. Luscombe. 2012. Evidence of non-random mutation rates suggests an evolutionary risk management strategy. *Nature* **485:** 95–98.

53. Gould, S.J. 1982. Darwinism and the expansion of evolutionary theory. *Science* **216:** 380–387.

54. King, D.G. & Y. Kashi. 2009. Heretical DNA sequences? *Science* **326:** 229–230.

55. Buschiazzo, E. & N.J. Gemmel. 2010. Conservation of human microsatellites across 450 million years of evolution. *Genome Biol. Evol* **2:** 153–165.

56. Riley, D.E. & J.N. Krieger. 2009. UTR dinucleotide simple sequence repeat evolution exhibits recurring patterns including regulatory sequence motif replacements. *Gene* **429:** 80–86.

57. Riley, D.E. & J.N. Krieger. 2009. Embryonic nervous system genes predominate in searches for dinucleotide simple sequence repeats flanked by conserved sequences. *Gene* **429:** 74–79.

58. Doyle, J., M. Csete & L. Caporale. 2006. An engineering perspective: the implicit protocols. In *The Implicit Genome.* Lynn H. Caporale, Ed.: pp. 294–298. Oxford University Press. New York.

Ann. N.Y. Acad. Sci. ISSN 0077-8923

ANNALS OF THE NEW YORK ACADEMY OF SCIENCES
Issue: *Effects of Genome Structure and Sequence on Variation and Evolution*

G4 motifs in human genes

Nancy Maizels

Departments of Immunology and Biochemistry, University of Washington, Seattle, Washington

Address for correspondence: Nancy Maizels, Departments of Immunology and Biochemistry, University of Washington, 1959 N.E. Pacific Street, H474A HSB, Seattle, WA 98195-7650. maizels@u.washington.edu

The G4 motif, $G_{\geq3}N_xG_{\geq3}N_xG_{\geq3}N_xG_{\geq3}$, is enriched in some genomic regions and depleted in others. This motif confers the ability to form an unusual four-stranded DNA structure, G4 DNA. G4 DNA is associated with genomic instability, which may explain depletion of G4 motifs from some genes and genomic regions. Conversely, G4 motifs are enriched downstream of transcription start sites, where they correlate with pausing. The uneven distribution of G4 motifs in the genome strongly suggests that mechanisms of selection act not only on one-dimensional genomic sequence, but also on structures formed by genomic DNA. The biological roles of G4 structures illustrate that, to understand genome function, it is important to consider the dynamic structural potential implicit in the G4 motif.

Keywords: DNA; G-quadruplex; repeat; replication; transcription

Introduction

Some sequence motifs recur in the genome, and others are depleted. What does this tell us about the genome, its evolution, and its functions? And what can we learn about the functions of specific motifs by studying their distribution and abundance in the genome? This paper focuses on the G4 motif, $G_{\geq3}N_xG_{\geq3}N_xG_{\geq3}N_xG_{\geq3}$, which confers an intrinsic potential to form structures other than canonical B form duplex DNA. The structure of DNA may vary from the canonical double helix, sometimes dramatically so, in a sequence-dependent manner. These diverse noncanonical structures can have significant effects on genomic stability and gene function. The G4 motif appears throughout the genome. Three chromosomal domains with specific functions are greatly enriched for G4 motifs ("G4^{hi}"): the telomeres (repeat TTAGGG); ribosomal DNA; and the immunoglobulin switch (S) regions, which are targets for the region-specific recombination essential to class switch recombination (see Kenter *et al.*[1]).

Although functionally important, G4 motifs can put the genome at risk because, as described later, replication through these structures can lead to genomic instability. Such instability has also been observed at other motifs with strong structure-forming potential. For example, trinucleotide repeats are a well-known example of deceptively simple sequences with strong structural potential. Many trinucleotide repeats can form stem–loop structures, which can be sites of sequence expansion or contraction. Expansions at unstable trinucleotide repeats contribute to at least 30 different neurodegenerative diseases.[2]

G4 motifs and G4 structures

The structures formed by regions of DNA bearing the G4 motif are referred to as G4 (or G-quadruplex) DNA. G4 structures can form in DNA or in RNA. G4 structures readily form *in vitro*; there is mounting evidence for the importance of G4 structures in biology.

The essential unit of a G4 structure is the G-quartet (Fig. 1A), in which four guanines interact by pairwise hydrogen bonds between the N7 of one guanine and the extracyclic amine of another.[3,4] Hydrogen bonds between guanines and stacking of the G-quartets upon one another confer very high thermodynamic stability upon G4 structures.

G4 structures are polymorphic

Nucleotides that do not participate in G-quartet formation (denoted as N_x in the consensus motif) form loops (Fig. 1B). Loop conformation is determined

doi: 10.1111/j.1749-6632.2012.06586.x

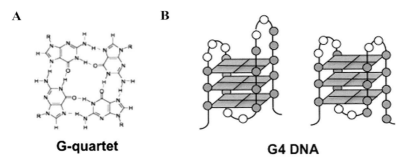

Figure 1. G-quartets and G4 DNA. (A) G-quartet formed by interactions among four guanine bases in DNA or RNA. (B) Intramolecular G4 DNA. The diagram illustrates two of the possible structures formed by the sequence $G_3N_3G_3N_2G_4N_2G_5$.

by local nucleotide sequence. Loops may extend over, under, or around the central stacked quartets to create a remarkable variety of structures.[5]

The icon typically used to represent G4 DNA does not give full credit to the potential structural diversity of even short sequences. Figure 1B shows a simple example of two of the numerous polymorphic structures that can be formed by the sequence $G_3N_3G_3N_2G_4N_2G_5$. The situation is undoubtedly more complex in the genome, where G4 motifs may contain more than four G-runs, and G-runs may be heterogeneous in length.

The potential for structural polymorphism makes it difficult to predict whether an individual G within a G4 motif will participate in G-quartet formation. Moreover, in order to form a G4 structure, the DNA duplex must undergo denaturation, as occurs in the course of transcription, replication, and recombination. These considerations suggest that G4 DNA structures are likely to differ from time to time within a cell and from cell to cell in a population. Structural polymorphism almost certainly explains the difficulty of identifying G-quartets by *in vivo* by DNA footprinting, as this technique pools information from populations of molecules. Structural polymorphism also means that standard sequence alignments overlook evolutionarily conserved G4 motifs embedded in a less conserved sequence unless algorithms have been designed to search for the G4 motif explicitly.

More than 300,000 G4 motifs can be identified in the human genome using a search algorithm that scores four or more adjacent runs of three or more guanines, in which loop size is constrained at 7 nt ($G_{\geq3}N_{1-7}G_{\geq3}N_{1-7}G_{\geq3}N_{1-7}G_{\geq3}$).[6] The relatively short length searched by this algorithm (on the

order of 30 nt) was chosen somewhat arbitrarily, and the distance over which G4 structures form *in vivo* may be longer.[7] Not surprisingly, the number of G4 motifs that can be identified in the genome increases with increasing loop length.

G4 structures form in single-stranded DNA

G4 structures form only when DNA is single-stranded. Transcription of regions bearing G4 motifs results in formation of characteristic structures, G-loops (Fig. 2), which contain a stable cotranscriptional RNA/DNA hybrid on the template strand and G4 DNA interspersed with single-stranded regions on the nontemplate strand.[8–10] G-loop and G4 DNA formation occur in only one transcriptional orientation, with a C-rich template strand. Transient denaturation that accompanies replication similarly enables formation of G4 DNA (Fig. 2).

G4 helicases unwind G4 DNA

G4 DNA that forms during replication is unwound by G4 helicases associated with the replication apparatus. These include the human helicases Bloom syndrome (BLM),[11] Werner syndrome (WRN),[12] and Fanconi anemia (FANCJ).[13,14] If G4 DNA is not resolved, it can lead to genomic instability.

The importance of replicative G4 DNA helicases has been dramatically demonstrated by experiments with model organisms. Nematodes lacking the FANCJ homolog *dog-1* exhibit genomic instability mapping to regions bearing G4 motifs.[15] *Saccharomyces cerevisiae* lacking the distinct G4 helicase Pif-1 exhibit instability at a G4[hi] reporter sequence, which is exacerbated by small molecule ligands selective for G4 structures.[16,17]

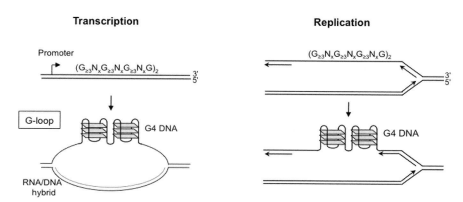

Figure 2. Formation of G4 DNA during transcription or replication. Right, transcription of two adjacent G4 motifs results in formation of a G-loop, containing a stable RNA/DNA hybrid (gray) and G4 DNA on the nontemplate strand. Left, G4 DNA forms on the template for lagging-strand replication.

Structure-function analysis has shown that the RecQ family helicases, such as BLM and WRN, bind to G4 DNA using a highly conserved domain found only in enzymes of this family.[18] This small RQC ("RecQ conserved") domain can bind G4 DNA independently of other domains of the enzyme, and with affinity in the nanomolar range. Comparable high affinity characterizes other protein/G4 DNA interactions. There has as yet been no structural analysis of protein bound to G4 DNA or to define the features of the G4 structure that are targets for G4 DNA–binding proteins.

G4 DNA causes instability in the human genome

Deficiencies in G4 helicases cause human diseases associated with genomic instability and cancer predisposition: BLM, WRN, and FANCJ.[19–23] Deficiency in a G4 helicase may have particular impact on a G4^{hi} chromosomal domain. BLM is characterized by immunodeficiency due to faulty class-switch recombination. Cells from individuals with WRN exhibit telomere shortening, which contributes to the premature aging characteristic of this disease.[24]

The potential for instability conferred by the G4 motif almost certainly accounts for the fact that some of the most polymorphic variable number tandem repeats in the human genome bear a G4 signature motif, including D4S43, GGGGAGGGGGAAGA; the insulin-linked hypervariable repeat, ACAGGGGTGTGGGG; MS1, AGGGTGGAG; MS32, CAGAATGGAGCAGGTGGC-CAGGGGTGACT; and CEB1, GGGGGGAGGG-AGGGTGGCCTGCGGAGGTCCCTGGGCTGA.

G4 motifs are selected in the human genome

The risk of genomic instability at G4 motifs suggests that selection might purge these motifs from functionally significant sequences, such as genes and their regulatory regions. However, this is not the case. We have shown that G4 motifs exhibit a very characteristic and highly uneven distribution among genes and gene regulatory regions.[25–28] This distribution argues for evolutionary selection of G4 motifs and for their likely importance in genomic function.

To score G4 motifs in human genes, we analyzed regions of sequence 100 nt in length in sliding windows of 100 nt, scoring each region that contained a G4 motif as a "hit," then quantifying the percentage of hits in the total numbers of windows searched.[25] We evaluated the distribution of gene ontology terms for each human gene across the spectrum of G4 DNA potentials. Most genes proved to be relatively depleted for G4 motifs ("G4^{lo}"), but some were considerably enriched ("G4^{hi}"). These differences in abundance of G4 motifs do not reflect coding capacity, but apply over entire genes, including both exons and introns.

G4 motifs correlate with gene function

Strikingly, we found that depletion or enrichment of G4 motifs correlates with gene function.[25] G4^{lo} genes function in nucleic acid binding, nucleosome assembly, ubiquitin-dependent proteolysis, cell adhesion, and regulation of cell division. In contrast, G4^{hi} genes function as transcription factors, growth factors, and cytokines, or in development, cell signaling, and muscle contraction. Tumor suppressor genes and oncogenes are at opposite ends of the spectrum: tumor suppressor genes are G4^{lo}, and oncogenes are G4^{hi}. The G4^{lo} status of tumor suppressor genes could be explained by selection for maximal genomic stability. But there is no obvious ready explanation for the G4^{hi} status of oncogenes.

Abundance of G4 motifs is independent of genomic environment

Our genomes consist of large segments of fairly homogeneous guanine + cytosine content, defined as isochores[29] (also see Ref. 30). Although the concept of isochores was put forward long before the sequence of the genome was known, it has proven to be robust.[31] We asked if abundance or depletion of G4 motifs might reflect gene position within larger isochores by computing the relative density of G4 motifs in a gene and the 20 kb of upstream and downstream flanking sequence.[25] Those calculations showed that, on average, tumor suppressor genes are reduced and oncogenes are enriched in G4 motifs, relative to their flanking sequences. Thus, at least for these classes of genes, distribution of G4 motifs within a gene correlates with gene function, but not with local genomic environment.

G4 structures can be targets of regulation

We have recently begun to appreciate that G4 motifs function in diverse and unanticipated contexts. Several examples in which G4 structures have been unambiguously correlated with gene regulation illustrate this functionality. A G4 DNA target controls antigen switching in *Neisseria gonorrhoeae*.[32] G4 DNA structures have been implicated in epigenetic regulation of human genes by ATRX, a chromatin remodelling factor deficient in an X-linked human genetic disease characterized by alpha thalassemia accompanied by mental retardation.[36] G4 RNA structures formed within transcripts of Fragile X genes that carry expansions of the trinucleotide repeat sequence CGG can titrate RNA processing factors to contribute to Fragile X disease pathology.[34] In addition, G4 structures have been shown to communicate changes in the cellular environment. G4 RNA structures formed within a 100 nt region of the p53 gene transcript determine 3′-processing of this mRNA upon induction of DNA damage.[7] This raises the possibility that a subset of G4 motifs may respond to and communicate stress genome-wide.

G4 motifs in regulatory regions of human genes

Figure 3 diagrams an idealized gene, with a regulatory region upstream of the transcription start site (TSS), and downstream exons and introns. We and others have shown that G4 motifs are enriched in the regions flanking the TSS.[26,35,36]

Strand-biased distribution of G4 motifs

The regions upstream and downstream of the TSS have very distinct potential to form nonduplex structures in the course of transcription. The upstream region may experience positive supercoiling that could be conducive to local formation of nonduplex structures. In contrast, the downstream region undergoes transient denaturation to enable passage of RNA polymerase. This prompted us to ask if there was evident strand bias of enrichment near the TSS. We found that upstream of the TSS, G4 motifs are comparably enriched on the template and nontemplate DNA strands.[26] Downstream of the TSS, G4 motifs are enriched on the nontemplate DNA strand.

To avoid overstating the case for G4 DNA in biology, it was important to ask if other mechanisms of regulation might explain G-richness flanking the TSS. We did this by masking sequence motifs for well-defined regulatory processes, such as CpG dinucleotides that are sites of methylation, and G-rich recognition motifs for common transcription factors, such as SP1 or KLF (Krüppel-like factor), and then recalculating distribution of G4 motifs.[26] Masking eliminated most of the G4^{hi} character of the region upstream of the TSS, suggesting that the G4^{hi} character of the region upstream of the TSS might reflect these well-defined processes.

The notion that regulatory factors might recognize G4 structures to control transcription had garnered considerable interest, particularly because many oncogene promoters contain at least one G4 motif upstream of the TSS. However, the modest residual G4 character remaining after masking suggested that duplex DNA was the predominant binding target. Nonetheless, there may be at least one exception to this rule. SP1 was recently shown to bind to G4 DNA with high affinity,[37] raising the possibility that SP1 might regulate gene expression by binding not only duplex DNA but also G4 DNA structures.

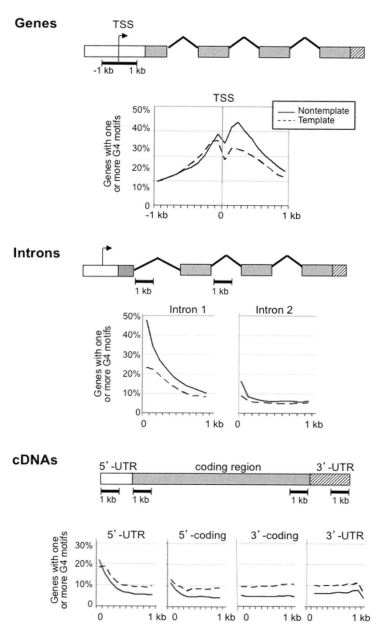

Figure 3. Distribution of G4 motifs in human genes. Percentage of genes with one or more G4 motifs per 100 bp interval was calculated for the nontemplate and template strands of the indicated regions flanking the transcription start site (TSS, top); at the 5′ ends of the first or second introns (middle); and at the 5′-ends of the 5′-UTR and coding region, or the 3′-ends of the coding region and 3′-UTR. Exons, shaded; 3′-UTR, hatched. Nontemplate and template strands are represented by solid and dashed lines, respectively. UTR, untranslated region.

The strand asymmetry of G4 motifs downstream of the TSS suggests that, in a fraction of genes, transcription of regions bearing G4 motifs may promote formation of nucleic acid structures that could serve as targets for regulation. Three kinds of unusual structures are associated with G-loop formation (Fig. 2): G4 DNA structures in the nontemplate strand of a G-loop, which may be recognized by regulatory factors; G4 structures in the pre-mRNA or mRNA, which may bind specific factors to determine mRNA processing or functions; and cotranscriptional RNA/DNA hybrids, which place special demands on RNA processing,[38,39] and may thus be subject to unusual regulation.

Conserved elements carrying G4 motifs in the first intron

The first intron represents a potentially privileged position within the genomic sequence, because it is very close to the TSS but does not appear in the mature mRNA or protein coding sequence. Proximity to the TSS would enable an element to load factors critical for regulation of gene expression, and enable their access to the core transcription complex. We found that nearly half of human genes carry one or more G4 motifs at the very 5′ end of the first intron, on the nontemplate DNA strand (Fig. 3, middle).[26] This enrichment displays clear strand bias, and is evident predominately on the nontemplate strand.

It is significant that G-rich intron 1 elements are conserved from humans to frogs in their potential to form G4 structures, even though the sequences themselves are not conserved. In many cases, these elements are quite long, comprising eight or more G-runs. Both these observations suggest that factors recognize structures formed in these regions. How and why remains to be discovered. This emphasizes, again, the importance of algorithms that search explicitly for G4 structures.

G4 motifs correlate with promoter-proximal pausing

RNA Pol II pauses just downstream of the promoter in many human genes. We showed that promoter-proximal pausing correlates with G-richness downstream of the promoter, and with G-rich intron 1 elements.[28] This raises the possibility that structures formed either in the DNA (by a pioneering round of transcription) or mRNA transcript may regulate pausing. This could occur either directly or via protein-mediated regulatory interactions. Pausing—and especially the release from pausing—may enable a rapid response to environmental or developmental signals, and thus provide another connection between genomic structures and the cellular response to stress.

Coding regions and UTRs are not enriched for G4 motifs

The enrichment of G4 motifs around the TSS and at the 5′ end of the first intron contrasts with the depletion and apparent selection against these motifs within coding regions. We found that 85% of human exons contain no G4 motifs.[27] A graph of distribution of G4 motifs within 1 kb downstream of ATG initiator codons and within 1 kb upstream of termination codons shows that relatively few genes carry these motifs, and where they do occur they predominate on the template rather than nontemplate DNA strand (Fig. 3, bottom). There are also few genes with G4 motifs in the 5′- and 3′-UTRs (note that there appears to be some enrichment for G4 motifs at the very 5′-end of the 5′-UTR and at the 5′-end of coding regions, but this can probably be ascribed to the contribution from the region just downstream of the TSS). The paucity of G4 motifs within genes provides further evidence for counterselection of these motifs within specific genomic regions.

For the future: what explains the distribution of G4 motifs in the genomic landscape?

The distinctive patterns of enrichment and depletion of G4 motifs support the view that these motifs have been selected during evolution and play key roles in genome biology. The genomic instability associated with G4 motifs can explain their depletion. Regulation by G4 structures may explain their enrichment, as suggested by the very intriguing examples that have come to light over the past few years. As we learn more about genomic regulation, more examples of function of these structures in regulation will undoubtedly be discovered. For example, systems analyses have the potential to correlate G4 motifs with gene expression under a variety of conditions, and thereby establish if and how G4 structures could exert global regulation in response to changing environmental conditions or cell stimuli. In addition, as genomewide analyses begin to systematically tally the spectrum of factors enriched at G4 motifs, we will learn whether G4 structures form rarely or frequently, and whether they are occasional or common regulatory targets.

The selection of G4 motifs in the genome highlights the importance of designing genomic search algorithms that look beyond primary DNA sequence to consider the dynamic structural potential of DNA.

Conflicts of interest

The author declares no conflicts of interest.

References

1. Kenter, A.L. , S. Feldman, R. Wuerffel, I. Achour, L. Wang & S. Kumar. 2012. Three dimensional architecture of the Igh locus facilitates class switch recombination. *Ann. N.Y. Acad. Sci.* **1267:** 86–94. This volume.
2. Mirkin, S.M. 2007. Expandable DNA repeats and human disease. *Nature* **447:** 932–940.
3. Gellert, M., M.N. Lipsett & D.R. Davies. 1962. Helix formation by guanylic acid. *Proc. Natl. Acad. Sci. U.S.A.* **48:** 2014–2018.
4. Sen, D. & W. Gilbert. 1988. Formation of parallel four-stranded complexes by guanine rich motifs in DNA and its implications for meiosis. *Nature* **334:** 364–366.
5. Phan, A.T., V. Kuryavyi & D.J. Patel. 2006. DNA architecture: from G to Z. *Curr. Opin. Struct. Biol.* **16:** 288–298.
6. Huppert, J.L. & S. Balasubramanian. 2005. Prevalence of quadruplexes in the human genome. *Nucleic Acids Res.* **33:** 2908–2916.
7. Decorsiere, A., A. Cayrel, S. Vagner & S. Millevoi. 2011. Essential role for the interaction between hnRNP H/F and a G quadruplex in maintaining p53 pre-mRNA 3′-end processing and function during DNA damage. *Genes Dev.* **25:** 220–225.
8. Duquette, M.L., P. Handa, J.A. Vincent, *et al.* 2004. Intracellular transcription of G-rich DNAs induces formation of G-loops, novel structures containing G4 DNA. *Genes Dev.* **18:** 618–1629.
9. Duquette, M.L., M.D. Huber & N. Maizels. 2007. G-rich proto-oncogenes are targeted for genomic instability in B-cell lymphomas. *Cancer Res.* **67:** 2586–2594.
10. Duquette, M.L., P. Pham, M.F. Goodman & N. Maizels. 2005. AID binds to transcription-induced structures in c-MYC that map to regions associated with translocation and hypermutation. *Oncogene* **24:** 5791–5798.
11. Sun, H., J.K. Karow, I.D. Hickson & N. Maizels. 1998. The Bloom's syndrome helicase unwinds G4 DNA. *J. Biol. Chem.* **273:** 27587–27592.
12. Fry, M. & L.A. Loeb. 1999. Human Werner syndrome DNA helicase unwinds tetrahelical structures of the fragile X syndrome repeat sequence d(CGG)n. *J. Biol. Chem.* **274:** 12797–12802.
13. London, T.B. *et al.* 2008. FANCJ is a structure-specific DNA helicase associated with the maintenance of genomic G/C tracts. *J. Biol. Chem.* **283:** 36132–36139.
14. Wu, Y., K. Shin-ya & R.M. Brosh, Jr. 2008. FANCJ helicase defective in Fanconia anemia and breast cancer unwinds G-quadruplex DNA to defend genomic stability. *Mol. Cell. Biol.* **28:** 4116–4128.
15. Kruisselbrink, E. *et al.* 2008. Mutagenic capacity of endogenous G4 DNA underlies genome instability in FANCJ-defective C. elegans. *Curr. Biol.* **18:** 900–905.
16. Ribeyre, C. *et al.* 2009. The yeast Pif1 helicase prevents genomic instability caused by G-quadruplex-forming CEB1 sequences *in vivo*. *PLoS Genet.* **5:** e1000475.
17. Piazza, A. *et al.* 2010. Genetic instability triggered by G-quadruplex interacting Phen-DC compounds in Saccharomyces cerevisiae. *Nucleic Acids Res.* **38:** 4337–4348.
18. Huber, M.D., M.L. Duquette, J.C. Shiels & N. Maizels. 2006. A conserved G4 DNA binding domain in RecQ family helicases. *J. Mol. Biol.* **358:** 1071–1080.
19. Sissi, C., B. Gatto & M. Palumbo. 2010. The evolving world of protein-G-quadruplex recognition: a medicinal chemist's perspective. *Biochimie* **93:** 1219–1230.
20. Monnat, R.J., Jr. 2010. Human RECQ helicases: roles in DNA metabolism, mutagenesis and cancer biology. *Semin. Cancer Biol.* **20:** 329–339.
21. Rossi, M.L., A.K. Ghosh & V.A. Bohr. 2010. Roles of Werner syndrome protein in protection of genome integrity. *DNA Repair (Amst)* **9:** 331–344.
22. Deans, A.J. & S.C. West. 2011. DNA interstrand crosslink repair and cancer. *Nat. Rev. Cancer* **11:** 467–480.
23. Deakyne, J.S. & A.V. Mazin. 2011. Fanconi anemia: at the crossroads of DNA repair. *Biochemistry (Mosc)* **76:** 36–48.
24. Crabbe, L., A. Jauch, C.M. Naeger, *et al.* 2007. Telomere dysfunction as a cause of genomic instability in Werner syndrome. *Proc. Natl. Acad. Sci. U.S.A.* **104:** 2205–2210.
25. Eddy, J. & N. Maizels. 2006. Gene function correlates with potential for G4 DNA formation in the human genome. *Nucleic Acids Res.* **34:** 3887–3896.
26. Eddy, J. & N. Maizels. 2008. Conserved elements with potential to form polymorphic G-quadruplex structures in the first intron of human genes. *Nucleic Acids Res.* **36:** 1321–1333.
27. Eddy, J. & N. Maizels. 2009. Selection for the G4 DNA motif at the 5′ end of human genes. *Mol. Carcinog.* **48:** 319–325.
28. Eddy, J. *et al.* 2011. G4 motifs correlate with promoter-proximal transcriptional pausing in human genes. *Nucleic Acids Res.* **39:** 4975–4983.
29. Bernardi, G. *et al.* 1985. The mosaic genome of warm-blooded vertebrates. *Science* **228:** 953–958.
30. Bernardi, G. 2012. The genome: an isochore ensemble and its evolution. *Ann. N.Y. Acad. Sci.* **1267:** 31–34. This volume.
31. Costantini, M., O. Clay, F. Auletta & G. Bernardi. 2006. An isochore map of human chromosomes. *Genome Res.* **16:** 536–541.
32. Cahoon, L.A. & H.S. Seifert. 2009. An alternative DNA structure is necessary for pilin antigenic variation in Neisseria gonorrhoeae. *Science* **325:** 764–767.
33. Law, M.J. *et al.* 2010. ATR-X syndrome protein targets tandem repeats and influences allele-specific expression in a size-dependent manner. *Cell* **143:** 367–378.
34. Santoro, M.R., S.M. Bray & S.T. Warren. 2011. Molecular mechanisms of fragile X syndrome: a twenty-year perspective. *Annu. Rev. Pathol.* **7:** 219–245.
35. Huppert, J.L. & S. Balasubramanian. 2007. G-quadruplexes in promoters throughout the human genome. *Nucleic Acids Res.* **35:** 406–413.
36. Zhao, Y., Z. Du & N. Li. 2007. Extensive selection for the enrichment of G4 DNA motifs in transcriptional regulatory

regions of warm blooded animals. *FEBS Lett.* **581:** 1951–1956.

37. Raiber, E.A., R. Kranaster, E. Lam, *et al.* 2011. A non-canonical DNA structure is a binding motif for the transcription factor SP1 *in vitro*. *Nucleic Acids Res.* **40:** 1499–1508.

38. Aguilera, A. 2005. mRNA processing and genomic instability. *Nat. Struct. Mol. Biol.* **12:** 737–738.

39. Gan, W. *et al.* 2011. R-loop-mediated genomic instability is caused by impairment of replication fork progression. *Genes Dev.* **25:** 2041–2056.

Ann. N.Y. Acad. Sci. ISSN 0077-8923

ANNALS OF THE NEW YORK ACADEMY OF SCIENCES

Issue: *Effects of Genome Structure and Sequence on Variation and Evolution*

Adaptive radiation of venomous marine snail lineages and the accelerated evolution of venom peptide genes

Baldomero M. Olivera,[1] Maren Watkins,[1] Pradip Bandyopadhyay,[1] Julita S. Imperial,[1] Edgar P. Heimer de la Cotera,[2] Manuel B. Aguilar,[2] Estuardo López Vera,[3] Gisela P. Concepcion,[4,5] and Arturo Lluisma[4,5]

[1]Biology Department, University of Utah, Salt Lake City, Utah. [2]Laboratorio de Neurofarmacología Marina, Departamento de Neurobiología Celular y Molecular, Instituto de Neurobiología, Universidad Nacional Autónoma de México, Campus Juriquilla, Mexico. [3]Instituto de Ciencias del Mar y Limnologia, Universidad Nacional Autónoma de México, Circuito Exterior s/n, Ciudad Universitaria, Coyoacán, Mexico. [4]Marine Science Institute, University of the Philippines, Diliman, Philippines. [5]Philippine Genome Center, University of the Philippines, Diliman, Philippines

Address for correspondence: Baldomero M. Olivera, Biology Department, University of Utah, 257 S. 1400 E., Salt Lake City, UT 84112. olivera@biology.utah.edu

An impressive biodiversity (>10,000 species) of marine snails (suborder Toxoglossa or superfamily Conoidea) have complex venoms, each containing approximately 100 biologically active, disulfide-rich peptides. In the genus *Conus*, the most intensively investigated toxoglossan lineage (~500 species), a small set of venom gene superfamilies undergo rapid sequence hyperdiversification within their mature toxin regions. Each major lineage of Toxoglossa has its own distinct set of venom gene superfamilies. Two recently identified venom gene superfamilies are expressed in the large Turridae clade, but not in *Conus*. Thus, as major venomous molluscan clades expand, a small set of lineage-specific venom gene superfamilies undergo accelerated evolution. The juxtaposition of extremely conserved signal sequences with hypervariable mature peptide regions is unprecedented and raises the possibility that in these gene superfamilies, the signal sequences are conserved as a result of an essential role they play in enabling rapid sequence evolution of the region of the gene that encodes the active toxin.

Keywords: venom peptides; accelerated evolution; Conidae; Turridae

Introduction

Venom has evolved independently many times in different phylogenetic lineages. The biodiversity of venomous animals that live in diverse habitats today may well exceed 10^5 species. Terrestrial groups such as snakes, spiders, scorpions, bees, and wasps are the most well known;[1,2] however, the marine environment has a comparable biodiversity of venomous animals, with those that frequently injure humans (such as jellyfish) being the most familiar.[3] Venomous marine snails, including the cone snails (genus *Conus*; family Conidae), constitute a significant fraction of the total biodiversity of all venomous animals (>10,000 species).[4] Two other groups of marine snails, the terebrids or auger snails (family Terebridae) and the turrids (family Turridae, s.l.) are venomous, and together with the cone snails,

traditionally constitute the suborder Toxoglossa (or alternatively, the superfamily Conoidea).

Marine snail venoms are complex

A noteworthy feature of marine snail venoms is their complexity; it was estimated, based on the initial biochemical purification of cone snail venoms, that each cone snail species has a repertoire of between 50–200 different venom components, mostly small disulfide-rich peptides.[5–8] Only a subset of these are expressed at one time in the venom of a single individual. However, on the basis of recent mass spectrometry analyses, it has been suggested that a much larger number of venom components are present, and that ~200 is a gross underestimate.[9] What has yet to be definitively established is the number of gene products encoded by distinct genes that are used in envenomation. Biass *et al.*[10] suggested that

doi: 10.1111/j.1749-6632.2012.06603.x

some of the >1,000 components they detected by mass spectrometry from *Conus consors* venom may be processing intermediates, or variants in the extent of posttranslational modification, and propeptides encoded by the same gene, and even degradation fragments that might be artifacts of venom collection and storage. Until the number of different genes encoding distinct peptides has been established, this issue remains unsettled. With the increasing number of transcriptome and genome analyses of cone snail venoms,[8,11–13] a determination of the number of venom peptides encoded by different genes used for envenomation should be forthcoming in the near future.

Cone snail venom peptides are examples of startlingly rapid evolution

A property of cone snail venom peptides that is now well established is their greatly accelerated evolution; essentially, each cone snail species has its own distinctive complement of venom peptides.

In general, only two or three gene superfamilies account for the majority of the peptides found expressed in the venom of each *Conus* species; however, which gene families predominate depends on the phylogenetic clade within the genus *Conus* to which that species belongs (see Fig. 1). In the past decade, the gene superfamilies that encode cone snail venom peptides have become increasingly well defined.[7] Most major gene superfamilies are found across the entire genus *Conus*, although a few seem to have a more restricted distribution.

The distribution of peptides in different gene superfamilies for the six *Conus* species that have yielded at least 30 different peptides from transcriptome analyses is shown in Figure 1A. Three of these are fish-hunting cone snails that belong to different clades: *C. bullatus* (*Textilia* clade), *C. consors* (*Pionoconus* clade), and *C. geographus* (*Gastridium* clade). One snail hunting species, *C. textile* (*Cylinder* clade) is shown, as well as two worm-hunting species, *C. litteratus* (*Strategeconus* clade) and *C. pulicarius* (*Puncticulis* clade).[16] Two species in Figure 1 have the greatest number of different peptides, *C. pulicarius* (82 sequences) and *C. geographus* (62 sequences). These peptides are from 16 different gene superfamilies; while some superfamilies are well represented in both (i.e., the O-superfamily[17,18]), peptides from seven of the superfamilies were found in only one of the two species.

Dramatic juxtapostion of conserved and hypervariable regions in *Conus* venom peptide gene superfamilies

A venom peptide is first translated as a larger precursor,[19,20] which can readily be divided into three regions: the signal sequence or preregion, an intervening or proregion, and, at the C-terminal end, the mature peptide region always present in single copy.[a] Throughout the genus, a superfamily retains the same conserved signal sequence at the N-terminus, juxtaposed against a hypervariable mature peptide (core) region at the C-terminus. With some exceptions, the number and arrangement of Cys residues is a conserved feature of all peptides in the same gene superfamily.[7]

Hypervariability of the biologically active toxin includes synonymous coding positions

In all major cone snail venom peptide gene superfamilies, there is a juxtaposition of an extremely conserved signal sequence region with the hypervariable mature toxin region; the C-terminal core sequences undergo accelerated evolution at an unprecedented rate. This striking feature was noted the very first time genes encoding *Conus* peptides were cloned.[19] It is important to note that the high rate of amino acid change that occurs in the core region is not merely because of extremely strong selection on individual toxin amino acids because the rate of change in synonymous codons is also greatly elevated.[22]

After undergoing posttranslational modification, the mature toxin peptide is released from the precursor by proteolytic cleavage. The posttranslationally modified mature peptide is the biologically active gene product injected into a targeted animal and binds specifically to a molecular target—an ion channel or receptor, and therefore, it would be this segment of the translated precursor, which is subject to intense diversifying selection. This is the

---

[a]Recently, it has been suggested that, for all posttranslationally modified but ribosomally synthesized peptides, no matter what their origin, the proregion be referred to as the "leader," and that the mature peptide region, which can be posttranslationally modified, as the "core peptide."[21]

A

Framework	Cys pattern	Superfamily	species: C.lit	C.tex	C.bul	C.pul	C.cons	C.geo
I	CC-C-C	A	3			10	14	12
III	CC-C-C-CC	M	8[a]	1	3	4	8	2
V	CC-CC	T	13			5	5	6
VI/VII	C-C-CC-C-C	O	8[b]	24	14	41[d]	15	18
VIII	C-C-C-C-C-C-C-C-C-C	S					3	5
IX	C-C-C-C-C-C	P	3	5		3	1	
XI	C-C-CC-CC-C-C	I				10[e]		
XIV	C-C-C-C	J, L	2		1	5		5
XV	C-C-CC-C-C-C-C	V	1[c]			1		
XVI	C-C-CC		1[c]					
XIX	C-C-C-CCC-C-C-C-C					1		
Conantokin	Non-disulfide-rich					3	1	2
Contryphan	Non-disulfide-rich		1		1	6		1
Contulakin	Non-disulfide-rich		2			1	4	1
Conkunitzin[f]	(Large)				1	2	7	1
Conophysin[f]	(Large)							1
Con-ikot-ikot[f]	(Large)							7

B

Figure 1. (A) Frequency of venom peptides. Comparison of relative frequency of conotoxin gene superfamilies in various Conus venom duct transcriptomes (Cys frameworks and superfamilies are indicated). (*C. lit*, *C. litteratus*),[14] (*C. tex*, *C. textile*),[15] (*C. bul*, *C. bullatus*),[8] (*C. pul*, *C. pulicarius*),[12] (*C. cons*, *C. consors*),[11] (*C. geo*, *C. geographus*).[13] [a] Includes sequences with three Cys residues; [b] includes sequences with two Cys residues; [c] the preregion (i.e., signal peptide) sequences indicate that this peptide belongs to a different superfamily; [d] includes O1 and some O3 superfamily sequences; [e] includes I1 and I2 sequences; [f] includes large peptides. (B) Shells of the cone snails analyzed above: (left to right) *C. textile*, *C. litteratus*, *C. geographus*, *C. bullatus*, *C. consors*, and *C. pulicarius*.

biological rationale offered for why hypermutation is focally localized to the mature toxin region.

The propeptide region is not hypervariable

The propeptide (leader) region seems to be conserved but undergoes a moderate degree of mutation; these leader sequences are much more conserved than the mature peptide but far less conserved than the adjoining N-terminal signal sequences. The propeptide regions contain binding sites for posttranslational modification enzymes, and thus are believed to facilitate the proper posttranslational processing of the core peptide region. Specifically, it has been demonstrated

for one posttranslational modification enzyme, which catalyzes the carboxylation of glutamate to γ-carboxyglutamate, that the recognition sequence within the propeptide region facilitates recruitment of the modification enzyme.[23] Evidence that disulfide isomerases bind to the propeptide has also been presented.[24]

The signal sequence is, unexpectedly, highly conserved

In contrast to every other known gene superfamily of secreted polypeptides, in *Conus* peptide gene superfamilies there is a striking sequence conservation of the signal sequence at the N-terminus of the precursor. This is unexpected because signal sequences do not require a specific amino acid sequence to function as secretion signals, but rather only need to have a general hydrophobic character. The unprecedented degree of sequence conservation strongly suggests that conopeptide gene superfamily signal sequences play additional important roles.

Specific examples of the juxtaposition of conserved and hypervariable regions are given in Figure 2. (Two groups of *Conus* venom peptides are shown: A-superfamily peptides, which generally have two disulfide linkages—the majority of those characterized are competitive nicotinic acetylcholine receptor antagonists; and S-superfamily peptides, which in contrast, have five disulfide bonds, and the molecular targets of these peptides are mostly unknown.)

The signal sequence and the portion of the mature toxin sequence encompassing the five central cysteine residues are magnified in the figure, and the corresponding nucleic acid sequences are shown in Figure 2B. This provides a particularly striking juxtaposition of absolute conservation in signal sequences, while in the interval of the toxin region shown, not a single amino acid is conserved except for the cysteine residues. Thus, the problem that remains, which has not been incisively addressed, is a mechanism of accelerated evolution resulting in almost absolute sequence conservation at the N-terminus, absent even a silent mutation, but an unprecedented rate of hypermutation at the C-terminal core region.

Rapid diversification of sequence enables rapid access to new prey

The data set on *Conus* venom peptides comprises what may be the most extensive set of rapidly diverging gene products for a single genus of animals. In some cases, thousands of different peptides in the same superfamily, from many different species, have been elucidated. The phylogenetic relationship between various *Conus* species has been determined, so that the pattern of divergence in a particular gene superfamily can be correlated with phylogenetic relationships between species (see, for example, Puillandre *et al.*).[25] For a substantial fraction of the peptides, there has been sufficient physiological characterization to allow the consequences of hypermutation to be assessed not only structurally but also functionally. An example of this type of

functional assessment is provided by the five peptides whose precursors are shown in Figure 2, and produced by the *Conus* species shown in Figure 3A.

Two gene superfamilies are represented, the A- and the S-superfamilies. The two S-superfamily peptide sequences shown in Figure 2 are the only examples in this superfamily where molecular targets have been identified: σ-conopeptide GVIIIA,[26] blocks the 5HT3 receptor (a ligand-gated ion channel that has serotonin as the native agonist), while αS-conopeptide RVIIIA[27] is a nicotinic receptor antagonist.

Functionally, the peptides shown fall into two classes: σ-conopeptide GVIIIA is an inhibitor of the serotonin 5HT3 receptor, whereas the other four, αS-conopeptide RVIIIA, α-conotoxins MII,[28] EI,[29] and CI[30] are nicotinic receptor antagonists (despite the fact that the first belongs to a different gene superfamily from the other three). Furthermore, α-conotoxin MII diverges functionally from the three other nicotinic antagonists in that it is targeted to a neuronal nicotinic receptor subtype (the α3β2 nicotinic receptor is the likely physiologically relevant target). The three other peptides all inhibit the muscle nicotinic receptor and paralyze fish.

Rapid convergent evolution across clades

Interestingly, there is an enormous sequence divergence between the latter three peptides although all come from fish-hunting cone snail venoms. There is compelling evidence[31] that the shift from an ancestral worm-hunting species to fish hunting took place independently in the new world to generate the *Chelyconus* clade (which includes *C. ermineus*, the source of α-conotoxin EI), and in the IndoPacific region (to generate the *Pionoconus* clade, which includes *C. catus*, the source of α-conotoxin CI). Although the selective pressure to evolve nicotinic receptor antagonists specifically targeted to the fish neuromuscular junction were presumably similar for the two peptides, because these peptides evolved from quite different worm-hunting ancestors, their evolutionary trajectories were very different, resulting in considerable divergence in their amino acid sequences, but convergence of their structures. Although it inhibits the same molecular target, the third peptide, αS-conotoxin RVIIIA is strikingly different structurally because an entirely different peptide superfamily was recruited for the same physiological purpose.

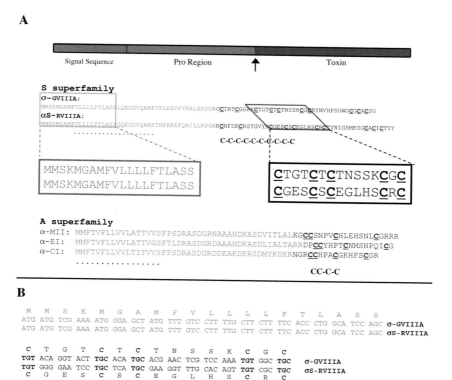

Figure 2. (A) Five precursor sequences from two Conus peptide gene superfamilies. The canonical organization of conopeptide precursors is shown diagrammatically, with the arrow indicating the proteolytic cleavage site that generates the mature toxin ("core") peptide. Two members of the S-superfamily are shown, with mature peptides containing five disulfide linkages (10 Cys residues). Three members of the A-superfamily are illustrated, with mature peptides that have two disulfide crosslinks (four Cys residues). The signal sequence and a segment of the mature toxins of the S-superfamily peptides have been magnified to illustrate the the total conservation in the signal sequence region, and the absence of any conservation of non-Cys amino acids in the mature toxin region. (B) The DNA sequences encoding the magnified aa sequences are shown; note the complete conservation of the nucleic acid sequences for the σ-GVIIIA and the αS-RVIII signal sequence regions; note the invariant cysteine codons. The molecular targets of these peptides are discussed in the text. The species from which the venom peptides are derived are illustrated in Figure 3A. Accession numbers: GVIIIA: FJ959110; RVIII: FJ959114; a-MII: P56636; a-EI: P50982; a-CI: FJ868069.1.

Accelerated evolution of venom peptides in other toxoglossan lineages

The vast majority of published studies on toxoglossan venoms focuses on cone snails, despite the fact that these snails are only a minor fraction of the total number of toxoglossan species. The venoms of most large toxoglossan clades have not been investigated at all; the only other group, apart from *Conus*, where a sufficient number of species have been examined so venom peptide evolution can be assessed is the family Turridae. Anatomical data and molecular phylogeny reveal that the classical family Turridae (s.l.) is a polyphyletic taxonomic assemblage, and there have been several proposals to break it up into multiple family groups. Many of the genera previously assigned to Turridae have now been transferred to other families, such as Drillidae and Raphitomidae. The total number of species in the family Turridae (s.s.), as presently restricted to a smaller number of genera by Bouchet *et al.*,[4] are probably comparable to the number of cone snail species. We discuss two superfamilies of turripeptides (as venom peptides from species in this family are designated).[32]

Turridae gene superfamilies are distinct from those expressed in cone snail venom

One issue is whether the major venom gene superfamilies of cone snails are also major components of the venoms of other venomous molluscs. At present, extensive transcriptome data have been published for only one species in the family

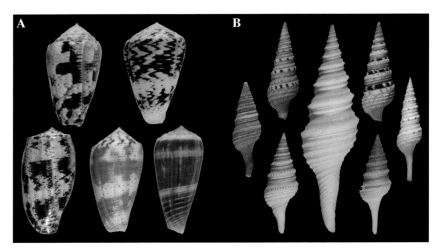

Figure 3. Shells of species analyzed in Figures 2 and 4. Shells are shown of cone snail and turrid species from which gene superfamily sequences discussed in this paper were obtained. The shells are not shown to scale. (A) Conus species from which conopeptide precursor sequences in Figure 2 were derived. Top row: *C. catus* (α-conotoxin CI); *C. ermineus* (α-conotoxin EI). Bottom row: *C. geographus* (σ-conotoxin GVIIIA); *C. magus* (α-conotoxin MII); and *C. radiatus* (αS-conotoxin RVIIIA). All of these species are believed to be fish hunters. *C. catus* and *C. magus* belong to the *Pionoconus* clade; *C. ermineus* to the *Chelyconus* clade, *C. geographus* to the *Gastridium* clade, and *C. radiatus* to the *Phasmoconus* clade. (B) Species in the family Turridae from which the sequences in Figure 4 are derived. Top row: *Xenuroturris cingulifera*; *X. olangoensis*. Large central shell, *Polystira albida*. Middle row: *Gemmula lisajoni*; *Lophiotoma albina*. Bottom row: *G. sogodensis*; *G. speciosa*. All of these species were collected in the Philippines, except for *P. albida*, which was collected from trawling in deep water in the Gulf of Mexico.

Turridae, *Lophiotoma olangoensis*.[33] *L. olangoensis* venom is comparable in complexity to cone snail venoms; some of the turripeptides in the venom are related to each other and belong to the same gene superfamily. However, a major conclusion was that there was little overlap between the gene superfamilies of cone snails, and the putative gene superfamilies that encode peptides expressed in *L. olangoensis* venom, suggesting that the expression of cone snail gene superfamilies does not generally extend to other major lineages of Toxoglossa.

Turripeptide superfamiliy expansion

An overview framework has emerged for how the diverse peptides in the venoms of the ~500 different *Conus* species have been generated.[6] As new species of cone[6,7] snails evolve, the type of focal hypermutation shown in Figure 2 occurs, primarily in a small subset of the gene superfamilies shown in Figure 1; the consequence is that the complement of peptides found in the venom of each particular *Conus* species diverges significantly from peptides in all other *Conus*. Does this hold true for other toxoglossan lineages as well?

Because the initial analysis of the *L. olangoensis* transcriptome, other species in the family Turridae have been analyzed,[34] so the similarities and differences in peptides from multiple species can be evaluated. Evidence for a gene superfamily subject to accelerated evolution was previously presented for a turripeptide gene superfamily expressed in the genus *Gemmula*.[34] More recent results demonstrate that a family of turripeptides that contain three disulfide bonds is expressed in four different genera of Turridae: the relevant precursor sequences are shown in Figure 4. It is clear that the polypeptides shown in Figure 4A belong to the same gene superfamily, given their highly conserved signal sequences. Although the mature toxin region is hypervariable, the arrangement of the six cysteine residues is conserved.

Gene superfamilies distributed across the entire Turridae undergo accelerated evolution

A preliminary molecular phylogeny of Turridae[35] has revealed that a basal clade in the family is the genus *Polystira*. This is a new world group, with several exceptionally large species found in both the Panamic region and the Caribbean. In the tree

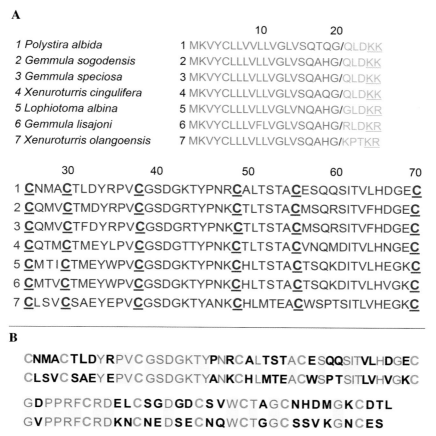

A

		10	20
1 *Polystira albida*	1	MKVYCLLVVLLVGLVSQTQG/	QLDKK
2 *Gemmula sogodensis*	2	MKVYCLLLVLLVGLVSQAHG/	QLDKK
3 *Gemmula speciosa*	3	MKVYCLLLVLLVGLVSQAHG/	QLDKK
4 *Xenuroturris cingulifera*	4	MKVYCLLLVLLVGLVSQAQG/	QLDKK
5 *Lophiotoma albina*	5	MKVYCLLLVLLVGLVNQAHG/	GLDKR
6 *Gemmula lisajoni*	6	MKVYCLLLVFLVGLVSQAHG/	RLDKR
7 *Xenuroturris olangoensis*	7	MKVYCLLLVLLVGLVSQAHG/	KPTKR

```
         30          40          50          60          70
1 CNMACTLDYRPVCGSDGKTYPNRCALTSTACESQQSITVLHDGEC
2 CQMVCTMDYRPVCGSDGRTYPNKCTLTSTACMSQRSITVFHDGEC
3 CQMVCTFDYRPVCGSDGRTYPNKCTLTSTACMSQRSITVFHDGEC
4 CQTMCTMEYLPVCGSDGTTYPNKCTLTSTACVNQMDITVLHNGEC
5 CMTICTMEYWPVCGSDGKTYPNKCHLTSTACTSQKDITVLHEGKC
6 CMTVCTMEYWPVCGSDGKTYPNKCHLTSTACTSQKDITVLHVGKC
7 CLSVCSAEYEPVCGSDGKTYANKCHLMTEACWSPTSITLVHEGKC
```

B

```
CNMACTLDYRPVCGSDGKTYPNRCALTSTACESQQSITVLHDGEC
CLSVCSAEYEPVCGSDGKTYANKCHLMTEACWSPTSITLVHVGKC

GDPPRFCRDELCSGDGDCSVWCTAGCNHDMGKCDTL
GVPPRFCRDKNCNEDSECNQWCTGGCSSVKGNCES
```

Figure 4. Turripeptide gene superfamilies. (A) Turripeptide sequences from seven species are aligned. All the sequences have a total length of 70 aa, with a conserved signal sequence of 20 aa. Predicted mature peptides have 3 disulfide linkages, in a P-like conopeptide framework (C–C–C–C—C—C). Note that compared to most conopeptide precursors, this turripeptide gene superfamily propeptide region is extremely short (and possibly even absent). (B) Two predicted mature peptide sequences from the superfamily shown in A, compared to two turripeptides from the Pg superfamily previously described by Heralde *et al.*[34] from two *Gemmula* species, *G. sogodensis*, and *G. kieneri*. Note that although both sets of peptides have six Cys residues, there is no alignment between the two groups, and these define two different P-like turripeptide superfamilies. In contrast to the sequences shown in A, the precursors of the Pg superfamily have a well-defined propeptide region of 23 aa.

presented by Heralde *et al.* this clade is the most distant from Indo-Pacific genera such as *Lophiotoma*, *Xenuroturris* and the Indo-Pacific branch of *Gemmula*. Because the genus *Polystira* is phylogenetically distant from the other groups, the presence of the precursor in *Polystira albida* venom shown in Figure 4A is strongly indicative that this gene superfamily is broadly expressed across the Turridae. Thus, as in *Conus*, there seem to be gene superfamilies distributed across the entire family Turridae that undergo accelerated evolution.

In Figure 4B, the predicted mature toxin from two peptides in the Pg turripeptide superfamily[34] are compared to two predicted mature peptides from the turripeptide superfamily in Figure 4A. Although both groups have a P-like conopeptide framework, they share no sequence homology with the P-conopeptide family, nor with each other. These are distinct from any peptides from *Conus* venoms.

Biological perspectives: radiation of biodiverse lineages

Lineage-specific acceleration of gene superfamily evolution

The biological implications of the molecular genetics discussed above are briefly discussed here. The data suggest that diversification of venom peptides in each major lineage of venomous toxoglossans occurs through the accelerated evolution of a few gene superfamilies that are characteristic of that lineage. There is an unusual conservation of signal sequences in such rapidly evolving gene superfamilies, compared to what is found in other families of secreted polypeptides. In all venom peptide precursors, the conserved signal sequences are strikingly juxtaposed against a rapidly evolving mature toxin region at the C-terminus, with only the cysteine codons conserved. A curious feature of the pattern of hypermutation, first pointed out by Conticello and Fainzilber,[15,36] is that within the hypervariable core region, not only are the cysteine residues conserved, but at each specific position, the codon encoding a Cys residue is invariant, with no mutational cycling between the two alternative Cys codons (see Fig. 2B for a specific example). It was suggested that mutations in the toxin region were introduced through the action of an error-prone DNA polymerase.

High conservation of signal sequence DNA suggests an undiscovered role

The data obtained so far suggest that although there is little overlap in the specific gene superfamilies expressed between the different major toxoglossan lineages, the extraordinary conservation of signal sequences is a general feature of toxoglossan venom peptide gene superfamilies. This sequence conservation strongly suggests additional functions besides being a standard secretion signal; two general types of hypotheses can be formulated.

Cell biological hypotheses: refined targeting

The first class of possibilities, which we call the cell biological hypotheses is that there may be greater specialization of secretory pathways in venom ducts compared to other tissues. Thus, if conopeptide or turripeptide precursors belonging to a particular gene superfamily were packaged with a specific subset of enzymes that promoted the correct folding and posttranslational modification of peptides in that superfamily, then the conserved sequences may not just direct a precursor polypeptide to pass through the endoplasmic reticulum, but may specify docking at a specific locus to package the precursor with specific chaperones, disulfide isomerase isoforms, modification enzymes, etc. This more refined targeting may require that signal sequences be recognized by sets of specific proteins. The essence of the hypothesis is that the signal sequence does not interact

just with the generic secretory system, but that additional recognition has evolved for each superfamily. This would provide positive selection for conserving specific amino acid sequences in each superfamily, although it would not explain the complete conservation of the DNA sequence as observed in Figure 2B. However, mechanisms similar to the conservation of sequence ribosomal RNA gene clusters could lead to DNA sequence conservation.

Genetic metacode hypotheses: functional constraints acting on the DNA

The other class, that we refer to as genetic metacode hypotheses is that the signal sequence region is an important determinant for modulating the rate of sequence change in the mature toxin segment. Thus, there may be some feature of the nucleic acid sequences encoding the signal sequence that increases the potential for hypermutation in the mature core sequence; one specific model would be to make the mature peptide region a preferred locus for double strand breaks or a hot spot for recombination. This type of additional function could account for the signal sequence conservation observed in some gene superfamilies even at the nucleic acid level (such as in the striking example of the absence of synonymous substitutions in the S-superfamily given in Fig. 2B). This could be rationalized by this type of hypothesis, first formulated by Caporale.[37] In one specific variation of these hypotheses, the requirement to be recognized by a biochemical mechanism that maintains the generation of hypermutability in the core peptide region is, paradoxically, the underlying explanation for signal sequence conservation. The two general classes of explanations are not mutually exclusive.

Lineage selection: favored by the rapid evolution of exogenes

As is discussed elsewhere,[38] venom genes are a readout of the biotic interactions of a particular toxoglossan species with its prey, predators, and competitors. Thus, as changes in the environment occur, novel ecological opportunities arise and new species can evolve. The ability of a lineage to opportunistically colonize new ecological niches as these arise depends on how quickly an ancestral species can become maximally fit for a new niche that becomes available. We suggest that a major component of the genetic change that accompanies speciation events in biodiverse lineages is optimization of biotic interactions for a new ecological niche by mutationally "tuning" those gene families that determine the fitness of a newly evolving species to deal with novel predators, prey, and competitors. Thus, in a specific lineage of animals, new species can be generated rapidly if the genes most relevant to success in a new ecological niche have the potential to be rapidly tuned to the new environment. In the venomous toxoglossa, this rapid tuning potential has apparently been achieved through the fascinating combination of extremely rapid sequence change in the mature toxin region accompanied by an almost complete suppression of base changes in the signal sequence region. The challenge is understanding what leads to this pattern—the availability of several complete toxoglossan genome sequences could clearly provide some useful insights.

As has been suggested elsewhere, toxoglossan venom peptide genes belong to a larger general class that we refer to as exogenes[38]—these are the genes that encode gene products that do not function endogenously, but that act on other organisms. We have suggested that the accelerated evolution of exogenes generally accompanies the rapid speciation that occurs as an explosive adaptive radiation generates a biodiverse lineage, such as has occurred for the major lineages of venomous marine snails.

Acknowledgments

The work described in this report was supported by NIH Grant GM48677 (to B.M.O.) and by ICBG Grant 1U01TW008163, Margo Haygood (PI), Oregon Health and Science University. This work was further supported by Grant IX211904 (to M.B.A.) and IN213808–3 (to E.L.V.) from the Programa de Apoyo a Proyectos de Investigación e Innovación Tecnológica, Universidad Nacional Autónoma de México (PAPIIT-UNAM), and Grants 75809 and 153915 (to E.L.V.) from the Consejo Nacional de Ciencia y Tecnología (CONACYT). We would like to thank Dr. Adolfo Gracia Gasca and the officers and crew of the RV Justo Sierra for their help in the trawling operations to obtain turrid snails during the oceanographic campaign COBERPES III. We thank Eric Schmidt for insightful discussions, and the stimulating and insightful suggestions of Lynn Caporale are also gratefully acknowledged.

Conflicts of interest

The authors declare no conflicts of interest.

References

1. Halstead, B.W. 1988. *Poisonous and Venomous Marine Animals of the World (Second Revised Edition)*. Princeton. The Darwin Press, Inc. New Jersey.

2. Tu, A.T., ed. 1984. *Handbook of Natural Toxins Volume 2. Insect Poisons, Allergens, and Other Invertebrate Venoms*. Marcel Dekker, Inc. New York, NY.

3. Tu, A.T., ed. 1988. *Handbook of Natural Toxins Volume 3. Marine Toxins and Venoms*. Marcel Dekker, Inc. New York, NY.

4. Bouchet, P. *et al.* 2011. A new operational classification of the Conoidea (Gastropoda). *J. Molluscan Stud.* **77:** 273–308.

5. Olivera, B.M. *et al.* 1985. Peptide neurotoxins from fish-hunting cone snails. *Science.* **230:** 1338–1343.

6. Olivera, B.M. 1997. *Conus* venom peptides, receptor and ion channel targets and drug design: 50 million years of neuropharmacology (E.E. Just Lecture, 1996). *Mol. Biol. Cell.* **8:** 2101–2109.

7. Terlau, H. & B.M. Olivera. 2004. *Conus* venoms: a rich source of novel ion channel-targeted peptides. *Physiol. Rev.* **84:** 41–68.

8. Hu, H. *et al.* 2011. Characterization of the Conus bullatus genome and its venom-duct transcriptome. *BMC Genom.* **12:** 60.

9. Davis, J., A. Alun Jones & R.J. Lewis. 2009. Remarkable inter- and intra-species complexity of conotoxins revealed by LC/MS. *Peptides.* **30:** 1222–1227.

10. Biass, D. *et al.* 2009. Comparative proteomic study of the venom of the piscivorous cone snail Conus consors. *J. Proteomics.* **72:** 210–218.

11. Terrat, Y. *et al.* 2012. High-resolution picture of a venom gland transcriptome: case study with the marine snail Conus consors. *Toxicon.* **59:** 34–46.

12. Lluisma, A.O. *et al.* 2012. Novel venom peptides from the cone snail Conus pulicarius discovered through next-generation sequencing of its venom duct transcriptome. *Marine Genom.* **5:** 43–51.

13. Hu, H. *et al.* 2012. Elucidation of the molecualr envenomation strategy of the cone snail Conus geographus through transcriptome sequencing of its venom duct. Submitted.

14. Pi, C. *et al.* 2006. Diversity and evolution of conotoxins based on gene expression profiling of *Conus litteratus*. *Genomics.* **88:** 809–819.

15. Conticello, S.G. *et al.* 2001. Mechanisms for evolving hypervariability: the case of conopeptides. *Mol. Biol. Evol.* **18:** 120–131.

16. Tucker, J.K. & M.J. Tenorio. 2009. Systematic Classification of Recent and Fossil Conoidean Gastropods: Conchbooks. Germany.

17. McIntosh, J.M. *et al.* 1995. A new family of conotoxins that blocks voltage-gated sodium channels. *J. Biol. Chem.* **270:** 16796–16802.

18. Olivera, B.M. 2002. *Conus* venom peptides: reflections from the biology of clades and species. *Annu. Rev. Ecol., Evol. System.* **33:** 25–42.

19. Woodward, S.R. *et al.* 1990. Constant and hypervariable regions in conotoxin propeptides. *EMBO J.* **1:** 1015–1020.

20. Colledge, C.J. *et al.* 1992. Precursor structure of ω-conotoxin GVIA determined from a cDNA clone. *Toxicon.* **30:** 1111–1116.

21. Van der Donk *et al.* Ribosomally synthesized and post-translationally modified peptide natural products, in Natural Product Reports. 2012. Accepted.

22. Olivera, B.M. *et al.* 1999. Speciation of cone snails and interspecific hyperdivergence of their venom peptides. Potential evolutionary significance of introns. *Ann. N.Y. Acad. Sci.* **870:** 223–237.

23. Bandyopadhyay, P.K. *et al.* 1998. Conantokin-G precursor and its role in γ-carboxylation by a vitamin K-dependent carboxylase from a *Conus* snail. *J. Biol. Chem.* **273:** 5447–5450.

24. Buczek, O., B.M. Olivera & G. Bulaj. 2004. Propeptide does not act as an intramolecular chaperone but facilitates protein disulfide isomerase-assisted folding of a conotoxin precursor. *Biochemistry.* **43:** 1093–1101.

25. Puillandre, N. *et al.* 2012. Molecular phylogeny, classification and evolution of conopeptides. *J. Mol. Evol.* Submitted.

26. England, L.J. *et al.* 1998. Inactivation of a serotonin-gated ion channel by a polypeptide toxin from marine snails. *Science.* **281:** 575–578.

27. Teichert, R.W., E.C. Jimenez & B.M. Olivera. 2005. αS-Conotoxin RVIIIA: a structurally unique conotoxin that broadly targets nicotinic acetylcholine receptors. *Biochemistry* **44:** 7897–7902.

28. Cartier, G.E. *et al.* 1996. α-Conotoxin MII (a-Ctx-MII) interaction with neuronal nicotinic acetylcholine receptors. *Soc. Neurosci. Abst.* **22:** 268.

29. Martinez, J.S. *et al.* 1995. α-Conotoxin EI, a new nicotinic acetylcholine receptor-targeted peptide. *Biochemistry.* **34:** 14519–14526.

30. Puillandre, N., M. Watkins & B.M. Olivera. 2010. Evolution of *Conus* peptide genes: duplication and positive selection in the A-superfamily. *J. Mol. Evol.* **70:** 190–202.

31. Imperial, J. *et al.* 2007. Using chemistry to reconstruct evolution: on the origins of fish-hunting in venomous cone snails. *Proc. Am. Philos. Soc.* **151:** 185–200.

32. Olivera, B.M., J. Imperial & G.P. Concepcion. 2012. Venom peptides from Conus and other Toxoglossan Marine Snails. In *Handbook of Biologically Active Peptides.* A.J. Kastin, Ed. Elsevier, Inc. Accepted.

33. Watkins, M., D.R. Hillyard & B.M. Olivera, 2006. Genes expressed in a turrid venom duct: divergence and similarity to conotoxins. *J. Mol. Evol.* **62:** 247–256.

34. Heralde, F.M., 3rd *et al.* 2008. A rapidly diverging superfamily of peptide toxins in venomous Gemmula species. *Toxicon.* **51:** 890–897.

35. Heralde, F.M., 3rd *et al.* 2010. The Indo-Pacific Gemmula species in the subfamily Turrinae: aspects of field distribution, molecular phylogeny, radular anatomy and feeding ecology. *Philip. Sci. Let.* **3:** 21–34.

36. Conticello, S.G. *et al.* 2000. Position-specific codon conservation in hypervariable gene families. *Trends Genet.* **16:** 57–59.

37. Caporale, L.H. 1984. Is there a higher level genetic code that directs evolution? *Mol. Cell. Biochem.* **64:** 5–13.

38. Olivera, B.M. 2006. Conus peptides: biodiversity-based discovery and exogenomics. *J. Biol. Chem.* **281:** 31173–31177.

Ann. N.Y. Acad. Sci. ISSN 0077-8923

ANNALS OF THE NEW YORK ACADEMY OF SCIENCES
Issue: *Effects of Genome Structure and Sequence on Variation and Evolution*

Integrons and gene cassettes: hotspots of diversity in bacterial genomes

Ruth M. Hall

School of Molecular Bioscience, The University of Sydney, NSW 2006, Australia

Address for correspondence: Ruth M. Hall, School of Molecular Bioscience, The University of Sydney, NSW 2006, Australia. ruth.hall@sydney.edu.au

Integrons are genetic units found in many bacterial species that are defined by their ability to capture small mobile elements called gene cassettes. Cassettes usually contain only one gene, potentially any gene, and an *attC* recombination site, and thousands of cassettes have been sequenced. A specialized IntI site–specific recombinase encoded by the integron recognizes *attC* and incorporates cassettes into an *attI* site located adjacent to the *intI* gene. Over 100 types of integrons have been found, most in bacterial chromosomes. They can all potentially share the same cassettes and, as recombination between *attC* in a cassette and an *attI* can occur repeatedly, an integron can contain from zero to hundreds of cassettes. Cassette arrays that are not located next to an *intI* gene, or solo cassettes at apparently random sites, are also seen. Hence, integrons contribute to generation of diversity in bacterial, plasmid, and transposon genomes and facilitate extensive sharing of information among bacteria.

Keywords: integron; gene cassettes; mobile elements; site-specific recombination; recombination site; antibiotic resistance

Gene cassettes and integrons: an overview

The well-documented role of gene cassettes in integrons found in transposons and on plasmids in the development of antibiotic resistance in Gram-negative bacteria attests to their potential as adaptive tools.[1–5] Integrons provide their hosts with natural cloning machinery that incorporates new genes in cassettes leading to the addition of genetic material to the repertoire of the bacterium that contains it.[1,2,4,5] The wide distribution of integrons in bacterial chromosomes indicates that they can play an important part in spreading information among bacteria in the same ecological niche. Integrons have been found in both terrestrial and marine bacterial communities[6] and have even been recovered from the extreme environment of hydrothermal vents.[7] Integrons have been very effective in exploiting bacterial genetic space and have been found in transposons, plasmids, and genomic islands as well as bacterial genomes. The expansion of the bacterial gene complement via gene cassette sharing can generate of diversity within members of a bacterial clone or species,[5,6,8] and the important role that a diverse collection of gene cassettes plays in enabling Gram-negative bacteria to overcome the deleterious effects of antibiotics[4,9] indicates that the integron–gene cassette system could contribute to survival in other specific or challenging environments. However, specific evidence for this in the form of cassette-carrying genes that have been proven to enhance survival in the particular environment is so far lacking.

Integrons: What are they?

The word *integron* was introduced just over 20 years ago to describe the first discovered element that was able to capture single genes via site-specific recombination.[1,10] Soon, a second integron type encoding a different integrase was recognized in the transposon Tn7, and then a third was found (reviewed in Ref. 4). Classes 1 and 3 were later shown to be associated with transposons or transposon remnants.[11,12] These distinct integron types, each encoding a different integrase, are now known as class 1, class 2, and class 3 integrons.[4] Members of each class have the same integrase but contain different gene cassettes, and, because they were recovered from antibiotic resistant clinical strains, all members of these three classes known at the time contained, predominantly but not exclusively, cassettes carrying antibiotic resistance genes.[1,2,4] A long evolutionary history for integrons was indicated by the fact that the IntI1, IntI2, and IntI3 proteins were not closely related—for example, IntI1 and IntI2 share 46% identity, and IntI1 and IntI3 share 60% identity.[13] Furthermore, in *Vibrio cholerae*, genes for diverse metabolic functions or pathogenicity determinants, such as lipocalins and toxins, had been shown to be associated with repeats called VCR (*V. cholerae* repeats). On further analysis,[14] VCR were found to have the properties of cassette-associated *attC* recombination sites (see Ref. 14 and gene cassettes below), indicating that there were gene cassettes in *V. cholerae,* and the first *V. cholerae* genome sequence revealed the presence of a further integron type (class 4) with a large array of gene cassettes.[8,15]

It was therefore not surprising that, over the last decade, these first four integron classes have been joined by over 100 integron classes (see, e.g., Refs. 8, 13, 16–21, and further references in Ref. 6). Each class includes a different *intI* gene (see Table 1 in Ref. 13 for an early compilation and comparison of IntIs). Integrons have been identified in genome sequences of a wide variety of bacterial species (see Ref. 6 for a detailed analysis and references) and in metagenomic data sets.[22] Others have been recovered experimentally from environmental samples— for example, Refs. 6, 7, and 23–25. More are likely to be discovered in the future as other environments are sampled.

Defining integrons and their properties

The term *integron* now describes a large family of genetic elements, all of which are able to capture members of the same family of small mobile elements, the gene cassettes. Though about a third of the potential integrons found in bacterial genomes did not include a gene cassette,[6] the role of integrons is to harbor and disseminate gene cassettes. To be able to do this, an integron needs two features, an *intI* gene and an *attI* recombination site (I for integron), and integrons are defined simply as *intI/attI* units that are or can be associated with gene cassettes.[1,6] Where cassettes are not found near a potential *intI* gene, experimental evidence that they

can be incorporated is needed before the existence of an integron is confirmed.

Generally, the *intI* gene and *attI* site are adjacent to one another, with the *attI* site upstream of *intI,* as shown in Figure 1. Occasionally, *attI* is downstream.[6,26] The *intI* genes encode a site-specific recombinase that belongs to a well-defined branch of the tyrosine recombinase or integrase family. Members of the IntI group are distinguished by a specific additional segment in the integrase[27] (labeled "IntI motif" in Fig. 1 of Ref. 13) that is important for recognition of *attC* sites. Integrons are classified using the sequence of the IntI recombinase, with members of the same class having the same (notionally >98% identity) integrase but variable cassette content. The *attI* recombination site adjacent to each *intI* gene is recognized by the IntI integrase, and each individual IntI exhibits strong specificity for its partner *attI* site and is unable to recognize *attI* sites that are paired with more distantly related members of the IntI family.[13,16,17]

The features required for a functional *attI* site other than a short, simple site sequence at the cassette insertion point cannot be deduced from examination of the DNA sequence—see, for example, Refs. 13, 23, 25, and 28. The architecture of these sites has been examined experimentally only for class 1[5,29–31] and class 3,[28] both of which include a simple site consisting of inversely oriented IntI binding sites and two further IntI binding sites, one that is necessary for site activity and one that enhances it. The additional sites may contribute to determining site polarity. More work is needed to determine how particular IntI recognize their own *attI* or closely related *attI* sites but are unable to see others.

Integrons differ from most other *int/att* units (e.g., integrating phage and integrative genomic islands) because they are not located within a discrete entity that they mobilize. Instead, integrons act *in trans* to mobilize and capture cassettes. The IntI integrases also recognize the *attC* recombination sites (C for cassette) of gene cassettes, enabling them to splice cassettes into the *attI* site (Fig. 1A). The *attI* × *attC* reaction is strongly preferred (see "Cassette movement" later), and new cassettes are added proximal to the integron as the first cassette. As this reaction is conservative and the *attI* site is reconstituted, it can occur over and over, creating a string of gene cassettes. Moreover, the high stringency exhibited by

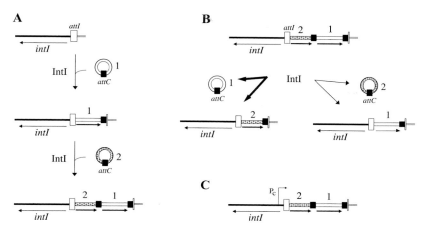

Figure 1. Integron structure and cassette movement. The integron is shown as a thin line with the *intI* gene marked with an arrow below, and the *attI* site is an open box. Cassettes are narrow open or stippled boxes with the *attC* site of the cassette a filled box and the gene shown as an arrow below. (A) Integration of gene cassettes. Two circular gene cassettes, 1 and 2, are shown recombining with an empty integron, then an integron containing one cassette. The final product, an integron containing two gene cassettes, is shown below. (B) Cassette excision. Excision of cassettes 1 and 2 to form the circular intermediate required for cassette integration and an integron with a single cassette is shown. The thickness of the arrows indicates the preference for *attC* × *attC* recombination. (C) The genes in cassettes are transcribed from the promoter P_c, shown located within the *intI* gene, as it is in the case of class 1 and class 2 integrons.

individual IntI with respect to their *attI* site does not apply to the cassette-associated recombination sites. Here, despite enormous diversity in the sequences of *attC* sites (see below), many different *attC* sites can be recognized by each different IntI. As a consequence, gene cassettes are readily exchangeable between different integron classes.[11,13,16,17,23,32–34] Thus, substantial diversity can be generated in any bacterium that harbors an integron via acquisition of novel gene cassettes.

In the integrons that have been studied so far, the number of cassettes varies from none (empty integrons) to hundreds in the case of *V. cholerae* and some other *Vibrio* species. Through the acquisition of different gene cassettes and of different numbers of gene cassettes, potentially in many different orders, enormous diversity can be generated in each class of integron. This is best illustrated by the diversity of gene cassettes containing antibiotic resistance genes (over 100 different ones) or other unknown or metabolic genes and the diversity of cassette combinations found in class 1 integrons.[2,4,9] The chromosomally located integrons in *Vibrio* species also have highly variable cassette arrays.[8]

When cassettes are present, they are viewed as part of the integron, and integrons are therefore composite structures made up of the integron backbone and an array of gene cassettes. However,

any individual integron–cassette combination can be described by class of the integron and a list of the cassettes in order. A discussion of the inappropriateness of the term *superintegron*, which seeks to include the number of cassettes in the definition of some integrons, including ones with relatively few gene cassettes, can be found elsewhere.[20,35] Classifying integrons by whether they contain cassettes with antibiotic resistance genes or by their location (chromosome, plasmid, transposon) or potential mobility is also problematic, as integrons of the same class can be found with or without resistance gene cassettes or in more than one context (see "Integrons on the move," below).

Gene cassettes: the mobile elements

Gene cassettes are the mobile elements in integron–cassette systems. They are among the simplest of known mobile element types but cannot move themselves, requiring the action in *trans* of an IntI.[2] Tracing the origins of gene cassettes has not yet been achieved; however, their ubiquity and abundance indicates that some organisms can create cassettes from single genes or gene pairs using pathways that have not been discovered so far. However, a comparison of sets of related cassettes revealed that, even when the DNA sequence divergence is over 20%, the gene and the *attC* are about equally diverged,[14] suggesting a very long evolutionary history for individual cassette families.

The defining feature of a cassette is the *attC* recombination site. Cassettes usually include only a single gene or open reading frame and an *attC* recombination site that allows them to be recognized by the IntI and, hence, is essential for mobility.[2] These sites were originally called 59-base elements (59-be) and subgroups are sometimes assigned additional names—for example, VCR. The architecture of the palindromic *attC* sites differs greatly from that of *attI* sites (see Refs. 5 and 32). The earliest studies recognized the enormous diversity in the sequences and lengths of *attC* sites, and observed that the salient feature of *attC* sites was the presence of moderately conserved, imperfect inverted repeats at the outer ends (reviewed in Refs. 2 and 14). A role for a folded structure was foreshadowed by DNA sequence comparisons that revealed that changes in the sequence at one end of an *attC* typically were reflected by changes at the other end that restored complementarity, and that the variation in the length of *attC* sites from 57 to over 140 bp was due to the presence of inverted repeats of different lengths and sequences in the center between the conserved portions.[36] Examination of all *attC* sites known in 1997 revealed relatively little complete sequence conservation but high conservation of complementarity between the left and right hand version of two sites that make up a simple site near the ends, and conservation of the additional bases in the left arm relative to the right.[37] However, a specific short DNA sequence (GTTRRRY) at the point of strand exchange is necessary for recombination to occur.[37] The features defined in this study continue to be used today to identify *attC* sites in linear DNA sequences, and hence form the basis of gene cassette identification.

The hypothesis that a folded DNA structure was involved was confirmed when the folded form of one strand (the bottom strand) was shown to be the recombinationally active *attC* site configuration.[38] The bases that are not matched by a base in the other half of an *attC* site bulge out from the double stranded part of a hairpin folded single strand (i.e., become extra helical) and removal of one of these additional bases or introduction of a complementary base inactivates the site.[5] The extrahelical bases interact with the IntI motif described earlier.[39,40] This interaction with a structural element (the bulged bases) rather than a specific DNA sequence is key to allowing *attC* sites with many different sequences to be recognized by IntIs. However, differences in the tolerance of individual IntI for the distance between the extrahelical bases has been reported recently,[40] and more needs to be learned about interactions of individual IntI with different *attC* sites.

Cassettes can potentially include any gene. Even in the earliest studies of antibiotic-resistant bacteria from the clinical situation, where a long period of antibiotic selection had been exerted, cassettes carrying open reading frames of unknown function were commonplace.[2,14,36] The function of a wider variety of genes found in cassettes has since been determined or deduced—for example, Refs. 8, 25, and 41–43—and cassettes that include potential antibiotic resistance genes are rare in integrons from sequenced genomes.[44]

With a few documented exceptions, a key feature of a cassette is that it contains only one gene. It is important to note that not all genes encode a protein and functionally important RNA molecules also need genes. Hence, cassette genes do not necessarily encode proteins.[6,42] Cassettes can contain two genes, particularly pairs of genes where one can cause damage if its action is not moderated by the second, such as restriction-modification enzymes or toxin-antidote pairs.[8,45] However, for the instances where more than one gene encoding unidentified polypeptides has been annotated in a cassette—for example, Ref. 26—the possibility that these represent pseudogenes cannot be eliminated. A further notable feature of genes in cassettes is that they almost always are in a specific orientation with respect to the *attC* site such that, when integrated into an integron, the 5′-end of the gene is nearest to the *attI* site. However, occasionally, genes have been found in the opposite orientation, including some of the identified toxin-antidote genes in the *V. cholerae* N16961 array.

Variants of gene cassettes that have lost most of the central part of the *attC* site,[2,29,46] effectively fusing the cassette to the next cassette or to the DNA segment next to it, have been found.[9,46] The fusion of two cassettes to form a single cassette means that the two genes will now move together; or if the cassette is at the last one in a cassette array, it will be immobilized.[46] An intriguing possibility is that such events may lead over time to the generation of stable new operons, although how genes with related functions would be brought into proximity other than by chance is not obvious. A model for how these reduced sites might arise involving the action of IntI has been proposed.[29] Though it is possible that cassettes containing only a reduced site may be excised at low frequencies compared to complete cassettes, available evidence does not support this.[46]

Expression of cassette genes

Most gene cassettes are compactly organized, and only rarely does a cassette include a promoter. Consequently, the majority of cassette-associated genes depend on the presence of an external, upstream promoter for their expression. A promoter, P_c, in the correct orientation has been proven to be present in the case of class 1,[4,47] class 2,[48] class 3[11] integrons and in an integron found in *Pseudomonas stutzeri*[34] (Fig. 1C). Thus, integrons can act as expression vectors for the genes captured in cassettes. Cassettes are integrated in a defined orientation, and usually in this orientation, the 5′-end of the gene is proximal to the P_c promoter supplied by the integron, as shown in Figure 1C. For class 1 integrons containing more than one cassette, it has been shown that genes in downstream cassettes are also transcribed from P_c[47] and the level of expression falls progressively for genes in downstream cassettes.[49] Consequently, a promoterless cassette may need to be located relatively close to P_c if its gene is to be expressed. However, some cassettes do include a promoter and even translational attenuation signals; for example,[50] the genes and orfs found in cassettes that are part of long cassette arrays may be expressed from promoters internal to or in the *attC* site of a preceding cassette(s). However, it is possible that in very long cassette arrays many downstream genes remain silent.

Gene cassette movement

Gene cassette capture (Fig. 1A) has been demonstrated experimentally in only a few cases.[11,17,34,51,52] Integration of circular gene cassettes into the *attI1* site of class 1 integrons[51] or the *attI3* site of class 3 integrons[11] was demonstrated using the IntI1 and IntI3 integrases, respectively, whereas integration of a cassette at an *attC* site was not observed. In a recent study, the frequencies of cassette incorporation were higher than those observed in earlier work and the stringency for the *attC* × *attI* reaction was confirmed.[52] The cassette

is incorporated precisely between the G and TT in one of the core sites (GTTrrry) of the simple site that is found in all *attI*.[37]

Gene cassettes can be excised from an integron to generate a free circular form.[53] Circular cassettes are the intermediates in cassette movement and can be reincorporated at an *attI* site. In contrast to integration, cassette excision is most efficient when the sites involved are both *attC* sites (Fig. 1B), and excision involving *attI* × *attC* recombination is at least an order of magnitude rarer.[13,16,17,46] These recombinational preferences, which are presumed to be a general feature of all integron classes, have led to the notion that cassettes go in at the front of a cassette array and "pop out" at the back. However, they can also presumably come out and reenter at the front, shuffling the order of cassettes in the array, as was seen in one study.[49]

It is known that integrons of different classes share cassettes because identical gene cassettes have been found in integrons from more than one class. For example, all of the cassettes that have been found in class 2 and class 3 integrons have also been found associated with class 1 integrons. There are many distinct groups of *attC* sites, though only a few of them have been clearly defined,[8,14] and integrons usually contain cassettes with a mixture of different *attC* types. Although the *attC* sites in the longer cassette arrays found in *Vibrio* chromosomal integrons are predominantly from a single *attC* group—for example, the VCR type for *V. choerae*—many other *attC* types are also found.[8] Thus, it appears that IntI-type integrases can all recognize a variety of cassette-associated *attC* sites, and this has been confirmed for several IntI (for example, see Refs. 11, 16, 17, 32, 36, and 40). However, the recombination efficiency of reactions involving different *attC* types varies, and individual IntI recognize some types of *attC* sites more efficiently than others. The broad recognition of *attC* types contrasts with the strong preference of each integrase for its own *attI* site.

Regulation of IntI expression

Recent studies have examined the regulation of *intI* gene expression, most often using the *intI1* gene of class 1 integrons. In class 1 and class 3 integrons, P_c is located within the *intI1* gene, which, as illustrated in Figure 1C, faces in the reverse direction, and hence transcripts from P_c run across the promoter for *intI1*.[11,47] The P_c promoter adversely affects *intI1* expression and hence recombination activity,[54,55] leading to an inverse correlation between promoter strength and the frequency of cassette excision.[54] The presence of binding sites for regulatory proteins LexA, FIS, IHF, and H-NS have been found in the vicinity of the *intI1* promoter, and LexA, FIS, and IHF can all repress expression of *intI1*.[55,56] The expression of IntI1 and other integron integrases, including IntI4, can also be induced as part of the SOS response after the DNA is damaged.[21,56] This may lead to the release of gene cassettes from damaged or dying cells into the environment, allowing them to be taken up by another host.

Solo cassettes or cassette arrays without integrons

Gene cassettes are not always associated with an integron. Cassette arrays without a nearby *intI* gene have been seen in some bacterial genomes, often ones that also carry an integron with a cassette array, such as *V. fischeri*[6] and *Saccharophagus degredans* (unpublished observations). In these cases, it seems likely that the adjacent sequence represents an *attI* site that has lost touch with its *intI* gene, but this needs to be established experimentally.

Recombination between an *attC* site and a secondary site has been shown to occur,[11,57] and this reaction can integrate a gene cassette at a location other than an *attI* site, enabling integration of new genes at many different positions in bacterial or plasmid genomes. However, when a promoterless cassette is incorporated at such a secondary site, the gene it contains will be expressed only if a promoter is present upstream. The IntI1 and IntI3 integrases, and probably other IntI, catalyze recombination with secondary sites that conform to a simple consensus (Ga/tT; see Refs. 11 and 57). Though this reaction is several orders of magnitude less efficient than recombination between *attC* and *attI,* it could contribute significantly to the evolution of plasmid and bacterial genomes, and a few cases where a known cassette carrying an antibiotic resistance gene is associated with a small plasmid[58–60] or with a bacterial chromosome[61] have been documented. In the future, computer-assisted searches are likely to reveal more such cassettes.

Integrons on the move

The phylogeny for the putative IntI found in 59 bacterial genome sequences did not always corre-late with the phylogeny for the *rpoB* gene in the same genomes, and this demonstrated clearly that integrons have frequently moved from one host to another and across large phylogenetic distances.[6] Moreover, within a particular species, some strains can include an integron while others do not. Whether other mobile elements, such as transposons or insertion sequences, are involved in the dissemination of integrons remains to be established.

A few cases of integrons of the same class in different organisms or locations have also been found. An integron in the genome of a *Pseudoalteromonas haloplanktis* strain encodes an IntI almost identical to one from an integron carrying an antibiotic resistance gene cassette found on a plasmid from *Vibrio salmonicida* (see Ref. 6). It seems likely that the class 1, class 2, and class 3 integrons associated with the recent dissemination of antibiotic resistance genes owe much of their capacity to do this to the fact that each has become associated with a transposon, thereby enhancing their ability to move onto plasmids and hence to move rapidly between organisms. In addition, there is evidence that both class 1 and class 3 integrons have been found in other locations in environmental bacteria.[62,63] The class 1 integrons were in a variety of contexts, indicating several transfer events.[62,63] However, the class 3 integrons found in *Deftia* species are in the same position in the bacterial chromosome.[63]

The spread of integrons among bacteria belonging to different phyla further emphasizes the capacity of integrons to spread information among bacteria. However, much remains to be learned. For example, why are so many putative integrons found in genome sequences not associated with gene cassettes? Are they integrons? Can they incorporate gene cassettes? Experimental studies are needed to answer these questions. In addition, how often and how easily are cassettes exchanged in the wild? Are genes in cassettes contributing to the survival of their host in challenging environments other than the presence of antibiotics?

Conflicts of interest

The author declares no conflicts of interest.

References

1. Hall, R.M. & C.M. Collis. 1995. Mobile gene cassettes and integrons: capture and spread of genes by site-specific recombination. *Mol. Microbiol.* **15:** 593–600.

2. Recchia, G.D. & R.M. Hall. 1995. Gene cassettes: a new class of mobile element. *Microbiology* **141:** 3015–3027.

3. Hall, R.M. 1997. Mobile gene cassettes and integrons: moving antibiotic resistance genes in Gram-negative bacteria. Symposium on antibiotic resistance: origins, evolution, selection and spread. *Ciba Foundation Symposium* **207:** 192–205.

4. Hall, R.M. & C.M. Collis. 1998. Antibiotic resistance in gram-negative bacteria: the role of gene cassettes and integrons. *Drug Resist. Updates* **1:** 109–119.

5. Hall, R.M. *et al.* 1999. Mobile gene cassettes and integrons in evolution. In *Molecular Strategies in Biological Evolution.* L.H. Caporale, Ed. *Ann. N.Y. Acad. Sci.* **870:** 68–80.

6. Boucher, Y. *et al.* 2007. Integrons: mobilizable platforms that promote genetic diversity in bacteria. *Trends Microbiol.* **15:** 301–309.

7. Elsaied, H. *et al.* 2007. Novel and diverse integron integrase genes and integron-like gene cassettes are prevalent in deep-sea hydrothermal vents. *Environ. Microbiol.* **9:** 2298–2312.

8. Rowe-Magnus, D.A. *et al.* 2003. Comparative analysis of superintegrons: engineering extensive genetic diversity in the Vibrionaceae. *Genome Res.* **13:** 428–442.

9. Partridge, S.R. *et al.* 2009. Gene cassettes and cassette arrays in mobile resistance integrons. *FEMS Microbiol. Rev.* **33:** 757–784.

10. Stokes, H.W. & R.M. Hall. 1989. A novel family of potentially mobile DNA elements encoding site-specific gene-integration functions: integrons. *Mol. Microbiol.* **3:** 1669–1683.

11. Collis, C.M. *et al.* 2002. Characterization of the class 3 integron and the site-specific recombination system it determines. *J. Bacteriol.* **184:** 3017–3026.

12. Partridge, S.R. *et al.* 2001. Family of class 1 integrons related to In4 from Tn*1696*. *Antimicrob. Agents Chemother.* **45:** 3014–3020.

13. Collis, C.M. *et al.* 2002. Integron-encoded IntI integrases preferentially recognize the adjacent cognate *attI* site in recombination with a 59-be site. *Mol. Microbiol.* **46:** 1415–1427.

14. Recchia, G.D. & R.M. Hall. 1997. Origins of the mobile gene cassettes found in integrons. *Trends Microbiol.* **5:** 389–394.

15. Heidelberg, J.F. *et al.* 2000. DNA sequence of both chromosomes of the cholera pathogen *Vibrio cholerae*. *Nature* **406:** 477–483.

16. Drouin, F., J. Mélançon & P.H. Roy. 2002. The IntI-like tyrosine recombinase of *Shewanella oneidensis* is active as an integron integrase. *J. Bacteriol.* **184:** 1811–1815.

17. Leon, G. & P.H. Roy. 2003. Excision and integration of cassettes by an integron integrase of *Nitrosomonas europaea*. *J. Bacteriol.* **185:** 2036–2041.

18. Holmes, A.J. *et al.* 2003. Recombination activity of a distinctive integron-gene cassette system associated with *Pseudomonas stutzeri* populations in soil. *J. Bacteriol.* **185:** 918–928.

19. Vaisvila, R. *et al.* 2001. Discovery and distribution of superintegrons among Pseudomonads. *Mol. Microbiol.* **42:** 587–601.

20. Hall, R.M. *et al.* 2007. What are superintegrons? *Nat. Rev. Microbiol.* **5:** 1–2.

21. Cambray, G. *et al.* 2011. Prevalence of SOS-mediated control of integron integrase expression as an adaptive trait of chromosomal and mobile integrons. *Mobile DNA* **2:** 6.

22. Nemergut, D.R. *et al.* 2008. Insights and inferences about integron evolution from genomic data. *BMC Genomics* **9:** 261.

23. Nield, B.S. *et al.* 2001. Recovery of new integron classes from environmental DNA. *FEMS Microbiol. Lett.* **195:** 59–65.

24. Nemergut, D.R., A.P. Martin & S.K. Schmidt. 2004. Integron diversity in heavy-metal-contaminated mine tailings and inferences about integron evolution. *Appl. Environ. Microbiol.* **70:** 1160–1168.

25. Elsaied, H. *et al.* 2011. Marine integrons containing novel integrase genes, attachment sites, attI, and associated gene cassettes in polluted sediments from Suez and Tokyo Bays. *ISME J.* **5:** 1162–1177.

26. Coleman, N. *et al.* 2004. An unusual integron in *Treponema denticola*. *Microbiology* **150:** 3524–3526.

27. Messier, N. & P.H. Roy. 2001. Integron integrases possess a unique additional domain necessary for activity. *J. Bacteriol.* **183:** 6699–6706.

28. Collis, C.M. & R.M. Hall. 2004. Comparison of the structure-activity relationships of the integron-associated recombination sites *attI3* and *attI1* reveals common features. *Microbiology* **150:** 1591–1601.

29. Partridge, S.R. *et al.* 2000. Definition of the *attI1* site of class 1 integrons. *Microbiology* **146:** 2855–2864.

30. Gravel, A., B. Fournier & P.H. Roy. 1998. DNA complexes obtained with the integron integrase IntI1 at the *attI1* site. *Nucleic Acids Res.* **26:** 4347–4355.

31. Collis, C.M. *et al.* 1998. Binding of the purified integron DNA integrase IntI1 to integron- and cassette-associated recombination sites. *Mol. Microbiol.* **29:** 477–490.

32. Collis, C.M. *et al.* 2001. Efficiency of recombination reactions catalyzed by class 1 integron integrase IntI1. *J. Bacteriol.* **183:** 2535–2542.

33. Hansson, K. *et al.* 2002. IntI2 integron integrase in Tn*7*. *J. Bacteriol.* **184:** 1712–1721.

34. Coleman, N.V. & A.J. Holmes. 2005. The native *Pseudomonas stutzeri* strain Q chromosomal integron can capture and express cassette-associated genes. *Microbiology* **151:** 1853–1864.

35. Hall, R.M. & H.W. Stokes. 2004. Integrons or super integrons? *Microbiology* **150:** 3–4.

36. Hall, R.M., D.E. Brookes & H.W. Stokes. 1991. Site-specific insertion of genes into integrons: role of the 59-base element and determination of the recombination cross-over point. *Mol. Microbiol.* **5:** 1941–1959.

37. Stokes, H.W. *et al.* 1997. Structure and function of 59-base element recombination sites associated with mobile gene cassettes. *Mol. Microbiol.* **26:** 731–745.

38. Johansson, C., M. Kamali-Moghaddam & L. Sundström. 2004. Integron integrase binds to bulged hairpin DNA. *Nucleic Acids Res.* **32:** 4033–4043.

39. MacDonald, D. *et al.* 2006. Structural basis for broad DNA-specificity in integron recombination. *Nature* **440:** 1157–1162.

40. Larouche, A. & P.H. Roy. 2011. Effect of *attC* structure on cassette excision by integron integrases. *Mobile DNA* **2:** 3.

41. Nield, B.S. *et al.* 2004. New enzymes from environmental cassette arrays: functional attributes of a phosphotransferase and an RNA-methyltransferase. *Protein Sci.* **13:** 1651–1659.

42. Holmes, A.J. *et al.* 2003. The gene cassette metagenome is a basic resource for bacterial genome evolution. *Environ. Microbiol.* **5:** 383–394.

43. Partridge, S.R. & R.M. Hall. 2005. Correctly identifying the streptothricin resistance gene cassette. *J. Clin. Microbiol.* **43:** 4298–4300.

44. Elbourne, L.D. & R.M. Hall. 2006. Gene cassette encoding a 3-N-aminoglycoside acetyltransferase in a chromosomal integron. *Antimicrob. Agents Chemother.* **50:** 2270–2271.

45. Pandey, D.P. & K. Gerdes. 2005. Toxin-antitoxin loci are highly abundant in free-living but lost from host-associated prokaryotes. *Nucleic Acids Res.* **33:** 966–976.

46. Partridge, S.R., C.M. Collis & R.M. Hall. 2002. Class 1 integron containing a new gene cassette, *aadA10*, associated with Tn*1404* from R151. *Antimicrob. Agents Chemother.* **46:** 2400–2408.

47. Collis, C.M. & R.M. Hall. 1995. Expression of antibiotic resistance genes in the integrated cassettes of integrons. *Antimicrob. Agents Chemother.* **39:** 155–162.

48. da Fonseca, É.L., F. dos Santos Freitas & A.C. Vicente. 2011. Pc promoter from class 2 integrons and the cassette transcription pattern it evokes. *J. Antimicrob. Chemother.* **66:** 797–801.

49. Collis, C.M. & R.M. Hall. 1992. Site-specific deletion and rearrangement of integron insert genes catalysed by the integron DNA integrase. *J. Bacteriol.* **174:** 1574–1585.

50. Stokes, H.W. & R.M. Hall. 1991. Sequence analysis of the inducible chloramphenicol resistance determinant in the Tn*1696* integron suggests regulation by translational attenuation. *Plasmid* **26:** 10–19.

51. Collis, C.M. *et al.* 1993. Site-specific insertion of gene cassettes into integrons. *Mol. Microbiol.* **9:** 41–52.

52. Gestal, A.M., E.F. Liew & N.V. Coleman. 2011. Natural transformation with synthetic gene cassettes: new tools for integron research and biotechnology. *Microbiology* **157:** 3349–3360.

53. Collis, C.M. & R.M. Hall. 1992. Gene cassettes from the insert region of integrons are excised as covalently closed circles. *Mol. Microbiol.* **6:** 2875–2885.

54. Jové, T. *et al.* 2010. Inverse correlation between promoter strength and excision activity in class 1 integrons. *PLoS Genet.* **6:** e1000793.

55. Cagle, C.A., J.E. Shearer & A.O. Summers. 2011. Regulation of the integrase and cassette promoters of the class 1 integron by nucleoid-associated proteins. *Microbiology* **157:** 2841–2853.

56. Guerin, E. *et al.* 2009. The SOS response controls integron recombination. *Science* **324:** 1034.

57. Recchia, G.D., H.W. Stokes & R.M. Hall. 1994. Characterisation of specific and secondary recombination sites recognised by the integron DNA integrase. *Nucleic Acids Res.* **22:** 2071–2078.

58. Recchia, G.D. & R.M. Hall. 1995. Plasmid evolution by acquistion of mobile gene cassettes: plasmid pIE723 contains the *aadB* gene cassette precisely inserted in a secondary site in the IncQ plasmid RSF1010. *Mol. Microbiol.* **15:** 179–187.

59. Ojo, K.K. *et al.* 2002. Identification of a complete *dfrA14* gene cassette integrated at a secondary site in a resistance plasmid of uropathogenic *Escherichia coli* from Nigeria. *Antimicrob. Agents Chemother.* **46:** 2054–2055.

60. Segal, H. & B.G. Elisha. 1999. Characterization of the *Acinetobacter* plasmid, pRAY, and the identification of regulatory sequences upstream of an *aadB* gene cassette on this plasmid. *Plasmid* **42:** 60–66.

61. Roy, P.H. *et al.* 2010. Complete genome sequence of the multiresistant taxonomic outlier Pseudomonas aeruginosa PA7. *PLoS One* **5:** e8842.

62. Gillings, M. *et al.* 2008. The evolution of class 1 integrons and the rise of antibiotic resistance. *J. Bacteriol.* **190:** 5095–5100.

63. Xu, H., J. Davies & V. Miao. 2007. Molecular characterization of class 3 integrons from *Delftia* spp. *J. Bacteriol.* **189:** 6276–6283.

Ann. N.Y. Acad. Sci. ISSN 0077-8923

ANNALS OF THE NEW YORK ACADEMY OF SCIENCES

Issue: *Effects of Genome Structure and Sequence on Variation and Evolution*

Creative deaminases, self-inflicted damage, and genome evolution

Silvestro G. Conticello

Core Research Laboratory, Istituto Toscano Tumori, Florence, Italy

Address for correspondence: Silvestro Conticello, Istituto Toscano Tumori, Core Research Laboratory, viale Pieraccini 6 Firenze, Toscana 50139, Italy. silvo.conticello@ittumori.it

Organisms minimize genetic damage through complex pathways of DNA repair. Yet a gene family—the AID/APOBECs—has evolved in vertebrates with the sole purpose of producing targeted damage in DNA/RNA molecules through cytosine deamination. They likely originated from deaminases involved in A>I editing in tRNAs. AID, the archetypal AID/APOBEC, is the trigger of the somatic diversification processes of the antibody genes. Its homologs may have been associated with the immune system even before the evolution of the antibody genes. The APOBEC3s, arising from duplication of AID, are involved in the restriction of exogenous/endogenous threats such as retroviruses and mobile elements. Another family member, APOBEC1, has (re)acquired the ability to target RNA while maintaining its ability to act on DNA. The AID/APOBECs have shaped the evolution of vertebrate genomes, but their ability to mutate nucleic acids is a double-edged sword: AID is a key player in lymphoproliferative diseases by triggering mutations and chromosomal translocations in B cells, and there is increasing evidence suggesting that other AID/APOBECs could be involved in cancer development as well.

Keywords: DNA editing; mutations; deamination; genome evolution

Introduction

The evolution of genes and genomes depends on the fine balance through which various cellular machineries handle the genetic information. Overall, most of the factors involved act to preserve the information to be passed to the next generation of cells and organisms, but exceptions exist. One of the most peculiar cases is represented by the AID/APOBECs, a gene family whose members are cytidine deaminases that induce targeted damage in nucleic acids. These enzymes arose at the beginning of the vertebrate radiation, likely from a group of deaminases involved in the A>I editing of the anticodon of tRNAs (Fig. 1).[1] The archetypal member of the family, activation-induced deaminase (AID), physiologically induces damage to DNA in the immunoglobulin loci, thus triggering all the processes that mediate the somatic diversification of the antibody genes.[2,3] Most of the other gene family members target nucleic acids in a diverse set of pathways: the APOBEC3 proteins are involved in a restriction path against retroviruses and mobile elements,[4] and APOBEC1 has been characterized as an RNA-editing enzyme.[5] On the other hand, much less is known with regard to the activity and functions of other family members such as APOBEC2, involved in muscle development.[6–8] Given the peculiarity of their roles, the activity of these DNA/RNA editors is tightly regulated, and their malfunction has been linked both to genetic diseases[9] and cancer.[10,11] While this gene family has been the object of intensive research, much has yet to be understood both in terms of the roles played by the AID/APOBECs, the regulation of their activity, and with regard to the way these enzymes may have affected the evolution of the vertebrate genomes.

From pyrimidine salvage to DNA editing

A first glimpse into the versatility of the AID/APOBECs is provided by the evolutionary trajectory that led to their appearance in the vertebrate genomes. In fact the AID/APOBECs are not unique

doi: 10.1111/j.1749-6632.2012.06614.x

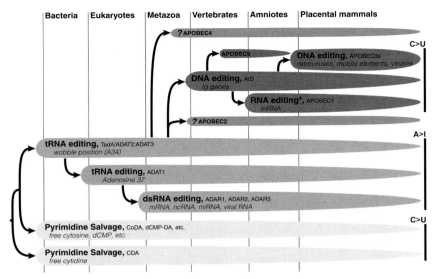

Figure 1. The evolutionary trajectory of the editing deaminases. The plot indicates the phylogenetic relations among the various zinc-dependent deaminases plotted against the time of their appearance. The various branches of the zinc-dependent deaminases are indicated in color: blue indicates A>I deamination; red indicates C>U deamination; and green indicates C>U deamination of free bases. The enzymes belonging to a given group are shown beside the known function; the physiologic targets are listed beneath. Branches of unknown catalytic activity are indicated only by the name of the protein group prefixed by a question mark (the APOBEC5 group, given its high similarity to AID and the APOBEC3s, is not labeled as unknown). APOBEC1 RNA editing is marked with an asterisk to indicate that APOBEC1 acts on DNA as well, but this activity has not yet been linked to a physiological role.

in their ability to induce modifications in (deoxy)nucleotides, but they are part of the much wider gene superfamily of the zinc-dependent deaminases acting on free cytosines and RNA-embedded adenines. These deaminases span the entire living kingdom and play a variety of roles. They all share a backbone comprising five beta strands and two alpha helices supporting the catalytic pocket.[1] Each zinc-dependent deaminase acquires its activity by specific add-ons at its ends or in the way the beta strands are connected.

Pyrimidine salvage

The zinc-dependent deaminases acting on free bases/nucleosides/nucleotides are involved in the pyrimidine salvage pathway and act as dimers/tetramers to deaminate cytosines into uracils.[12,13] Despite their widespread presence, their role in the salvage pathway is not essential, as some organisms lack some of these enzymes, and with their function served by alternative pathways: for example, neither cytosine deaminases nor their activity have been identified so far in any metazoa,[14] while dCTP deamination in *E. coli* is accomplished

through an enzyme that is not related to the zinc-dependent deaminases.[15]

RNA editing

tRNA. More intriguing is the path taken by another branch of the zinc-dependent deaminases, those targeting purines instead of pyrimidines. These enzymes—the TadA/ADAT2—deaminate adenosine 34 at the wobble position of the anticodon in several tRNAs.[16] The resulting inosine enables the edited tRNA to recognize more than one codon. While bacterial TadA deaminases act as homodimers, the eukaryotic ones form heterodimers with catalytically inactive relatives, the ADAT3 proteins. These are the earliest enzymes known to modify what was up to then a fixed set of genetic information. While it is not possible to infer which branch of the zinc-dependent deaminases came first, it is notable that the efficiency of the degenerate genetic code also depends on the activity of these enzymes. In fact there are no known organisms lacking them and their ablation is lethal.[16,17] The TadA/ADAT2s have supplied the backbone for the evolution of all known A>I RNA editing enzymes. Closer relatives are the Tad1p/ADAT1, eukaryotic deaminases, which specifically target tRNAs at the adenosine 37 position.[18]

mRNA, microRNA, and other functions. The other A>I editing deaminases, the ADARs, originated in animals from the ADAT1s.[19] They constitute a group of related proteins in which recognition of the substrate—contrary to all other nucleic acids deaminases—is mediated by a series of double-strand RNA-binding domains.[20] They were originally characterized for their involvement in the regulation of brain function through A>I editing that changed the amino acid specified by a codon. Later research showed that these proteins affect many other dsRNA targets in a number of ways. ADAR proteins are thus involved in the regulation of coding and noncoding RNAs, in the splicing of mRNA by creating novel splice sites, and in the modulation of the processing of miRNA.[20] Moreover, they might be involved in viral restriction,[21] but their activity has been highjacked by viruses such as HIV to their advantage.[22]

Evolutionary origin of DNA editing

It is from the tRNA-editing branch of the zinc-dependent deaminases that the AID/APOBECs most likely originated.[23] Note that AID/APOBECs, which mainly target cytosines in the context of DNA, seem to originate from a group of deaminases targeting adenosines in the context of RNA rather than from one of the pyrimidine deaminases acting on free nucleobases. This might be due to the characteristics of target recognition: it could have been easier to slightly modify the catalytic pocket to accommodate a cytidine rather than develop *ex novo* the structure supporting the recognition and binding of the polynucleotide chain. In this respect, it is very intriguing the finding that ADAT2 from trypanosomes *in vitro* acts as a bifunctional enzyme, being able to perform both A>I editing in tRNAs and C>U deamination in bacteria.[24]

Regulated DNA editing

AID: editing immunoglobulin genes

AID was identified after a long quest to understand the mechanistic processes behind the somatic hypermutation and class switch recombination processes of the antibody genes during the antigen-driven response in activated B cells.[9,25,26] It was named based on its similarity to zinc-dependent deaminases and

on its expression pattern in activated B cells. AID can be found in both bony and cartilaginous fish (e.g., Ref. 27), which are the earliest vertebrate to possess an adaptive immune system based on antibody genes. The evolution of AID is tightly connected to that of the vertebrate immune system. The most remarkable signature of such coevolution is the sequence preference context of AID (each AID/APOBEC presents specific preferences for the sequence context in which the targeted cytidine is located): the complementarity determining regions (CDRs), the portions of the antibody genes more targeted by the somatic hypermutation, are enriched in sequences specific for AID (WRC, where W is A or T and R is A or G).[28] Similarly, the switch regions—the molecular targets of the class switch recombination—are formed by repetitive elements bearing the AID recognition sequence (see Kenter *et al.*[66] in this volume). Using mutants of AID that target a different sequence preference context significantly changes the resulting mutation pattern on antibody genes.[29]

AID-like molecules predate the vertebrate adaptive immune response

Bona fide AID homologs have been found in jawless fish such as lamprey.[30] In these organisms, the immune system is based on variable lymphocyte receptors (VLRs), a class of receptors composed of leucin-rich repeats (LRR),[31] which somatically diversify through gene conversion (beyond immune diversification, gene conversion is actively used during meiotic recombination, see Cole *et al.*[64] in this volume for a discussion). It has been hypothesized that these early AID homologs could trigger the diversification in a manner similar to how AID acts in vertebrates such as chicken: targeted AID-dependent DNA damage triggers the homologous recombination pathways to repair the VLRs using nearby LRR units as DNA templates. This would imply that the AID/APOBECs were associated with immunity even before the appearance of the immunoglobulin genes, and even before the appearance of the RAG genes, the enzymes that mediate the primary diversification of the antibody gene in vertebrates. If this hypothesis proves true, it would mean that AID-like molecules not only were part of the machinery to diversify the effectors of the immune system, but also were essential in defining the framework itself underlying their diversification. Thus, while initially linked to the primary diversification of the immune receptors, AID moved to the secondary diversification of the antibody genes once more effective machinery—the RAG genes—took over their original role. Finally, given the observation that a hypermutation-based machinery might exist in the immune system of the *Biomphalaria* snails, the story could even go further back, especially if some of the deaminase-encoding genes identified in molluscans share the ability to edit DNA.[32,33]

Antiretroviral role of nucleic acid editing enzymes

The evolution of machinery able to target nucleic acids provided the basis for yet another role: the ability of the AID/APOBECs to induce DNA damage could be used to protect the genome from rogue nucleic acids such as viral genomes. The first evidence for such a role came with the identification of APOBEC3G as the factor involved in the restriction of HIV in the absence of Vif, an HIV accessory gene.[34] APOBEC3G is physiologically loaded into the HIV virions and affects the reverse transcription by targeting the nascent DNA genome (for example, Ref. 35). However, the HIV protein Vif counteracts this defense mechanism by targeting APOBEC3G for proteasomal degradation.[36] While ineffective against HIV due to HIV Vif, APOBEC3G has been shown to successfully restrict other retroviruses, including SIV. Though SIV contains Vif, it is, however, ineffective against human APOBEC3G (for example, Ref. 37). The acquisition of the antiretroviral function led the AID/APOBECs into an arms race that caused the rapid expansion of the ancestral APOBEC3 locus in placental mammals.[38–41] Many species have evolved different arrays of APOBEC3 genes, with the primates in particular having the largest one, comprising seven APOBEC3s. Beyond HIV, APOBEC3 proteins physiologically restrict other retroviruses, such as the Friend virus and AKV murine leukemia virus in mice.[42–44] *In vitro* experiments show many APOBEC3s as inhibiting factors in the propagation of a number of retroviruses and viruses, such as SIV, HTLV-1, HBV, and AAV.[45,46] And there are indications that they could be part of a more general pathway involved in the removal of foreign DNA.[47]

Restriction of transposable elements

Analogous to its restriction of exogenous retroviruses, APOBEC3s can also restrict various classes of transposable elements.[45] In such a role, the AID/APOBECs might police the genome by limiting the jumping of retrotransposons. In fact, their mutagenic signature—a bias toward G>A changes can be found in the relics of past (retro)transpositional events.[48]

AID/APOBECs and genome evolution

It is intriguing that, while the APOBEC3s are restricted to placental mammals, other derivatives of AID—not yet characterized—are present in other tetrapod lineages.[33] It could thus be possible that, within the boundaries of the AID/APOBECs, several generations of molecules could have evolved to fulfill similar functions in an arms race against viruses and mobile elements, and that the sudden expansion of the APOBEC3 genes in primates might just represent their adaptation to primate-specific threats (see Ref. 65 for a discussion on exogenes). Thus, genomes are constantly remodeled by specific interactions with genomic parasites—viruses and mobile elements. But these interactions depend both on the nature of the attacking factors and on the response that a given cell/organism can exhibit. In such a response, the AID/APOBECs could play a major role: the differences we observe among species in content and type of genomic junk could depend on the presence in each given species of AID/APOBEC molecules with different efficiencies and specificities in contrasting the genomic threats.

Back to RNA editing

APOBEC1 merits separate mention as the first among the AID/APOBECs to be identified and alone among them in being characterized as an RNA-editing enzyme. APOBEC1 is part of the editing complex that deaminates C6666 to U in human apolipoprotein B (ApoB), a structural component of lipid transport.[49,50] The resulting premature stop codon causes the translation of a truncated form of ApoB, which is important in the assembly of chylomicrons. Since this initial discovery, additional mRNA targets have been discovered[51,52] and other possible roles have been attributed to APOBEC1: regulating the stability of specific mRNAs;[52,53] controlling DNA methylation;[54,55] and being part of an innate restriction pathway against retroviruses, similar to that of the APOBEC3s (see, for example, Refs. 56–58). Moreover, similar to the other AID/APOBECs, APOBEC1 was also shown to mutate DNA in bacteria, though this activity was considered to be a relic from its evolution.[59]

The past and the present of APOBEC1 function

Until recently, APOBEC1 was thought to have originated from a duplication of the AID locus in mammals, but the availability of genome sequences from a variety of organisms led to the identification of APOBEC1-like molecules before the divergence of the amniotes.[33] Whereas this could have represented just an evolutionary tidbit, the analysis of reptilian APOBEC1 provides insight into a few aspects that could be relevant to understanding the role of APOBEC1. Thus, APOBEC1 is not involved in the editing of reptilian ApoB, suggesting that such a role is a later acquisition in mammals. Moreover, the reptilian APOBEC1 is able to deaminate DNA, as can the other AID/APOBECs. Its mutational context preference is very similar to that of the mammalian homolog in spite of the time of divergence. This is remarkable, as more conserved APOBECs, such as the APOBEC3s, deaminate DNA with different sequence contexts.[1] Moreover, the sequence context at the RNA editing site is independent from APOBEC1, as the recognition of the mRNA targets rely on other factors of the editing complex. All this suggests that a selective pressure acts on APOBEC1 similar to that which allowed the coevolution of AID mutational context with its target sequences in the Ig locus. Thus, the ability of APOBEC1 to target DNA might not be a remnant from the past, superseded by its novel function as an RNA-editing enzyme: its ability to target DNA in a specific context could be relevant for another physiological role yet to be uncovered.

Cancer connection

The AID/APOBECs exert their activity in diverse physiological contexts, and their abilities have effectively shaped the evolution of the vertebrates. But their ability to mutate nucleic acids comes at a steep price: the mutagenic action of AID plays a major role in the onset of the genetic alterations characteristic of mature B cell tumors.[60,61] In addition, in the case of APOBEC1, there is a connection between its expression and the onset of cancer. There is also evidence supporting the ability of APOBEC3 proteins to cause DNA damage in the cells expressing them.[62,63] It is clear that a fine balance sustains the beneficial functions of the AID/APOBECs: small changes—through misregulation or concurrent DNA damage—can tip them toward an aberrant role. Ultimately, this underscores the role these molecules have played in the evolution of our genome.

Summary

DNA editing—the process of regulated damage to specific DNA targets—has evolved in vertebrates as a means to efficiently diversify the immune repertoire. Its effectors, the AID/APOBEC deaminases, have been incorporated in a variety of cellular processes. Their origin is likely from deaminases involved in A>I editing in tRNAs: adaptation of the catalytic site to deaminate deoxycytosine to deoxyuracil, combined with the ability to target nucleic acids, has provided the basis for the rise of these DNA-editing enzymes. As an intended or unintended consequence of their activity, the AID/APOBECs have effectively shaped the evolution of the vertebrate genomes both as a direct cause of mutations and genetic alterations and as indirect factors, due to their ability to target endogenous and exogenous DNA and RNA molecules.

Acknowledgment

This work was supported by a start-up grant from the Istituto Toscano Tumori, and by a grant of the Italian Ministry of Health (GR-2008-1141464).

Conflicts of interest

The author declares no conflicts of interest.

References

1. Conticello, S.G., M.A. Langlois, Z. Yang & M.S. Neuberger. 2007. DNA deamination in immunity: AID in the context of its APOBEC relatives. *Adv. Immunol.* **94:** 37–73.
2. Di Noia, J.M. & M.S. Neuberger. 2007. Molecular mechanisms of antibody somatic hypermutation. *Annu. Rev. Biochem.* **76:** 1–22.
3. Kracker, S. & A. Durandy. 2011. Insights into the B cell specific process of immunoglobulin class switch recombination. *Immunol. Lett.* **138:** 97–103.
4. Malim, M.H. 2009. APOBEC proteins and intrinsic resistance to HIV-1 infection. *Philos. Trans. R. Soc. Lond. B. Biol. Sci.* **364:** 675–687.
5. Blanc, V. & N.O. Davidson. 2010. APOBEC-1-mediated RNA editing. *Wiley Interdiscip. Rev. Syst. Biol. Med.* **2:** 594–602.

6. Liao, W., S.H. Hong, B.H. Chan, *et al.* 1999. APOBEC-2, a cardiac- and skeletal muscle-specific member of the cytidine deaminase supergene family. *Biochem. Biophys. Res. Commun.* **260:** 398–404.

7. Anant, S., J.O. Henderson, D. Mukhopadhyay, *et al.* 2001. Novel role for RNA-binding protein CUGBP2 in mammalian RNA editing. CUGBP2 modulates C to U editing of apolipoprotein B mRNA by interacting with apobec-1 and ACF, the apobec-1 complementation factor. *J. Biol. Chem.* **276:** 47338–47351.

8. Sato, Y., H.C. Probst, R. Tatsumi, *et al.* 2010. Deficiency in APOBEC2 leads to a shift in muscle fiber type, diminished body mass, and myopathy. *J. Biol. Chem.* **285:** 7111–7118.

9. Revy, P., T. Muto, Y. Levy, F. Geissmann, *et al.* 2000. Activation-induced cytidine deaminase (AID) deficiency causes the autosomal recessive form of the Hyper-IgM syndrome (HIGM2). *Cell* **102:** 565–575.

10. Yamanaka, S., M.E. Balestra, L.D. Ferrell, *et al.* 1995. Apolipoprotein B mRNA-editing protein induces hepatocellular carcinoma and dysplasia in transgenic animals. *PNAS* **92:** 8483–8487.

11. Okazaki, I.M., A. Kotani & T. Honjo. 2007. Role of AID in tumorigenesis. *Adv. Immunol.* **94:** 245–273.

12. Song, B.H. & J. Neuhard. 1989. Chromosomal location, cloning and nucleotide sequence of the Bacillus subtilis cdd gene encoding cytidine/deoxycytidine deaminase. *Mol. Gen. Genet.* **216:** 462–468.

13. Betts, L., S. Xiang, S.A. Short, *et al.* 1994. Cytidine deaminase. The 2.3 A crystal structure of an enzyme: transition-state analog complex. *J. Mol. Biol.* **235:** 635–656.

14. Nishiyama, T., Y. Kawamura, K. Kawamoto, *et al.* 1985. Antineoplastic effects in rats of 5-fluorocytosine in combination with cytosine deaminase capsules. *Cancer Res.* **45:** 1753–1761.

15. Johansson, E., M. Fanø, J.H. Bynck, *et al.* 2005. Structures of dCTP deaminase from Escherichia coli with bound substrate and product: reaction mechanism and determinants of mono- and bifunctionality for a family of enzymes. *J. Biol. Chem.* **280:** 3051–3059.

16. Gerber, A.P. & W. Keller. 1999. An adenosine deaminase that generates inosine at the wobble position of tRNAs. *Science* **286:** 1146–1149.

17. Wolf, J., A.P. Gerber & W. Keller. 2002. tadA, an essential tRNA-specific adenosine deaminase from Escherichia coli. *EMBO J.* **21:** 3841–3851.

18. Gerber, A., H. Grosjean, T. Melcher & W. Keller. 1998. Tad1p, a yeast tRNA-specific adenosine deaminase, is related to the mammalian pre-mRNA editing enzymes ADAR1 and ADAR2. *EMBO J.* **17:** 4780–4789.

19. Gerber, A.P. & W. Keller. 2001. RNA editing by base deamination: more enzymes, more targets, new mysteries. *Trends Biochem. Sci.* **26:** 376–384.

20. Nishikura, K. 2010. Functions and regulation of RNA editing by ADAR deaminases. *Annu. Rev. Biochem.* **79:** 321–349.

21. Taylor, D.R., M. Puig, M.E. Darnell, *et al.* 2005. New antiviral pathway that mediates hepatitis C virus replicon interferon sensitivity through ADAR1. *J. Virol.* **79:** 6291–6298.

22. Doria, M., F. Neri, A. Gallo, *et al.* 2009. Editing of HIV-1 RNA by the double-stranded RNA deaminase ADAR1 stimulates viral infection. *Nucleic Acids Res.* **37:** 5848–5858.

23. Conticello, S.G. 2008. The AID/APOBEC family of nucleic acid mutators. *Genome Biol.* **9:** 229.

24. Rubio, M.A., I. Pastar, K.W. Gaston, *et al.* 2007. An adenosine-to-inosine tRNA-editing enzyme that can perform C-to-U deamination of DNA. *PNAS* **104:** 7821–7826.

25. Muramatsu, M., V.S. Sankaranand, S. Anant, *et al.* 1999. Specific expression of activation-induced cytidine deaminase (AID), a novel member of the RNA-editing deaminase family in germinal center B cells. *J. Biol. Chem.* **274:** 18470–18476.

26. Muramatsu, M., K. Kinoshita, S. Fagarasan, *et al.* 2000. Class switch recombination and hypermutation require activation-induced cytidine deaminase (AID), a potential RNA editing enzyme. *Cell* **102:** 553–563.

27. Conticello, S.G., C.J. Thomas, S.K. Petersen-Mahrt & M.S. Neuberger. 2005. Evolution of the AID/APOBEC family of polynucleotide (deoxy)cytidine deaminases. *Mol. Biol. Evol.* **22:** 367–377.

28. Wagner, S.D., C. Milstein & M.S. Neuberger. 1995. Codon bias targets mutation. *Nature* **376:** 732.

29. Wang, M., C. Rada & M.S. Neuberger. 2010. Altering the spectrum of immunoglobulin V gene somatic hypermutation by modifying the active site of AID. *J. Exp. Med.* **207:** 141–153, S1–S6.

30. Rogozin, I.B., L.M. Iyer, L. Liang, *et al.* 2007. Evolution and diversification of lamprey antigen receptors: evidence for involvement of an AID-APOBEC family cytosine deaminase. *Nat. Immunol.* **8:** 647–656.

31. Boehm, T., N. McCurley, Y. Sutoh, *et al.* 2011. VLR-based adaptive immunity. *Annu. Rev. Immunol.* **30:** 203–220.

32. Zhang, S.M., C.M. Adema, T.B. Kepler & E.S. Loker. 2004. Diversification of Ig superfamily genes in an invertebrate. *Science* **305:** 251–254.

33. Severi, F., A. Chicca & S.G. Conticello. 2011. Analysis of reptilian APOBEC1 suggests that RNA editing may not be its ancestral function. *Mol. Biol. Evol.* **28:** 1125–1129.

34. Sheehy, A.M., N.C. Gaddis, J.D. Choi & M.H. Malim. 2002. Isolation of a human gene that inhibits HIV-1 infection and is suppressed by the viral Vif protein. *Nature* **418:** 646–650.

35. Harris, R.S. & M.T. Liddament. 2004. Retroviral restriction by APOBEC proteins. *Nat. Rev. Immunol.* **4:** 868–877.

36. Niewiadomska, A.M. & X.F. Yu. 2010. Host Restriction of HIV-1 by APOBEC3 and viral evasion through Vif. *Curr. Top. Microbiol. Immunol.* **339:** 1–25.

37. Mariani, R., D. Chen, B. Schröfelbauer, *et al.* 2003. Species-specific exclusion of APOBEC3G from HIV-1 virions by Vif. *Cell* **114:** 21–31.

38. Sawyer, S.L., M. Emerman & H.S. Malik. 2004. Ancient adaptive evolution of the primate antiviral DNA-editing enzyme APOBEC3G. *PLoS Biol.* **2:** E275.

39. Zhang, N., G. Wu, H. Wu, *et al.* 2004. Purification, characterization and sequence determination of BmKK4, a novel potassium channel blocker from Chinese scorpion Buthus martensi Karsch. *Peptides* **25:** 951–957.

40. Larue, R.S., S.R. Jonsson, K.A. Silverstein, *et al.* 2008. The artiodactyl APOBEC3 innate immune repertoire shows evidence for a multi-functional domain organization that

existed in the ancestor of placental mammals. *BMC Mol. Biol.* **9:** 104.

41. LaRue, R.S., V. Andrésdóttir, Y. Blanchard, S.G. Conticello, *et al.* 2009. Guidelines for naming nonprimate APOBEC3 genes and proteins. *J. Virol.* **83:** 494–497.

42. Santiago, M.L., M. Montano, R. Benitez, *et al.* 2008. Apobec3 encodes Rfv3, a gene influencing neutralizing antibody control of retrovirus infection. *Science* **321:** 1343–1346.

43. Takeda, E., S. Tsuji-Kawahara, M. Sakamoto, *et al.* 2008. Mouse APOBEC3 restricts Friend leukemia virus infection and pathogenesis in vivo. *J. Virol.* **82:** 10998–11008.

44. Langlois, M.A., K. Kemmerich, C. Rada & M.S. Neuberger. 2009. The AKV murine leukemia virus is restricted and hypermutated by mouse APOBEC3. *J. Virol.* **83:** 11550–11559.

45. Chiu, Y.L. & W.C. Greene. 2008. The APOBEC3 cytidine deaminases: an innate defensive network opposing exogenous retroviruses and endogenous retroelements. *Annu. Rev. Immunol.* **26:** 317–353.

46. Chen, H., C.E. Lilley, Q. Yu, *et al.* 2006. APOBEC3A is a potent inhibitor of adeno-associated virus and retrotransposons. *Curr. Biol.* **16:** 480–485.

47. Stenglein, M.D., M.B. Burns, M. Li, *et al.* 2010. APOBEC3 proteins mediate the clearance of foreign DNA from human cells. *Nat. Struct. Mol. Biol.* **17:** 222–229.

48. Esnault, C., O. Heidmann, F. Delebecque, *et al.* 2005. APOBEC3G cytidine deaminase inhibits retrotransposition of endogenous retroviruses. *Nature* **433:** 430–433.

49. Navaratnam, N., J.R. Morrison, S. Bhattacharya, *et al.* 1993. The p27 catalytic subunit of the apolipoprotein B mRNA editing enzyme is a cytidine deaminase. *J. Biol. Chem.* **268:** 20709–20712.

50. Teng, B., C.F. Burant & N.O. Davidson. 1993. Molecular cloning of an apolipoprotein B messenger RNA editing protein. *Science* **260:** 1816–1819.

51. Skuse, G.R., A.J. Cappione, M. Sowden, *et al.* 1996. The neurofibromatosis type I messenger RNA undergoes base-modification RNA editing. *Nucleic Acids Res.* **24:** 478–485.

52. Rosenberg, B.R., C.E. Hamilton, M.M. Mwangi, *et al.* 2011. Transcriptome-wide sequencing reveals numerous APOBEC1 mRNA-editing targets in transcript 3′ UTRs. *Nat. Struct. Mol. Biol.* **18:** 230–236.

53. Anant, S. & N.O. Davidson. 2000. An AU-rich sequence element (UUUN[A/U]U) downstream of the edited C in apolipoprotein B mRNA is a high-affinity binding site for Apobec-1: binding of Apobec-1 to this motif in the 3′

untranslated region of c-myc increases mRNA stability. *Mol. Cell Biol.* **20:** 1982–1992.

54. Morgan, H.D., W. Dean, H.A. Coker, *et al.* 2004. Activation-induced cytidine deaminase deaminates 5-methylcytosine in DNA and is expressed in pluripotent tissues: implications for epigenetic reprogramming. *J. Biol. Chem.* **279:** 52353–52360.

55. Guo, J.U., Y. Su, C. Zhong, *et al.* 2011. Hydroxylation of 5-methylcytosine by TET1 promotes active DNA demethylation in the adult brain. *Cell* **145:** 423–434.

56. Petit, V., D. Guétard, M. Renard, *et al.* 2009. Murine APOBEC1 is a powerful mutator of retroviral and cellular RNA in vitro and in vivo. *J. Mol. Biol.* **385:** 65–78.

57. Cervantes Gonzalez, M., R. Suspene, M. Henry, *et al.* 2009. Human APOBEC1 cytidine deaminase edits HBV DNA. *Retrovirology* **6:** 96.

58. Gee, P., Y. Ando, H. Kitayama, *et al.* 2011. APOBEC1-mediated editing and attenuation of herpes simplex virus 1 DNA indicate that neurons have an antiviral role during herpes simplex encephalitis. *J. Virol.* **85:** 9726–9736.

59. Harris, R.S., S.K. Petersen-Mahrt & M.S. Neuberger. 2002. RNA editing enzyme APOBEC1 and some of its homologs can act as DNA mutators. *Mol. Cell* **10:** 1247–1253.

60. Ramiro, A., B.R. San-Martin, K. McBride, *et al.* 2007. The role of activation-induced deaminase in antibody diversification and chromosome translocations. *Adv. Immunol.* **94:** 75–107.

61. Pasqualucci, L., G. Bhagat, M. Jankovic, *et al.* 2008. AID is required for germinal center-derived lymphomagenesis. *Nat. Genet.* **40:** 108–112.

62. Landry, S., I. Narvaiza, D.C. Linfesty & M.D. Weitzman. 2011. APOBEC3A can activate the DNA damage response and cause cell-cycle arrest. *EMBO Rep.* **12:** 444–450.

63. Suspène, R., M.M. Aynaud, D. Guétard, *et al.* 2011. Somatic hypermutation of human mitochondrial and nuclear DNA by APOBEC3 cytidine deaminases, a pathway for DNA catabolism. *PNAS* **108:** 4858–4863.

64. Cole, F., S. Keeney & M. Jasin. 2012. Preaching about the converted: how meiotic gene conversion influences genomic diversity. *Ann. N.Y. Acad. Sci.* **1267:** 95–102. This volume.

65. Olivera, B.M., M. Watkins, P. Bandyopadhyay, *et al.* 2012. Adaptive radiation of venomous marine snail lineages and the accelerated evolution of venom peptide genes. *Ann. N.Y. Acad. Sci.* **1267:** 61–70. This volume.

66. Kenter, A.L., S. Feldman, R. Wuerffel, *et al.* 2012. Three-dimensional architecture of the IgH locus facilitates class switch recombination. *Ann. N.Y. Acad. Sci.* **1267:** 86–94. This volume.

Ann. N.Y. Acad. Sci. ISSN 0077-8923

ANNALS OF THE NEW YORK ACADEMY OF SCIENCES
Issue: *Effects of Genome Structure and Sequence on Variation and Evolution*

Three-dimensional architecture of the IgH locus facilitates class switch recombination

Amy L. Kenter, Scott Feldman, Robert Wuerffel, Ikbel Achour, Lili Wang,[*] and Satyendra Kumar

Department of Microbiology and Immunology, University of Illinois College of Medicine, Chicago, Illinois

Address for correspondence: Amy L. Kenter, Department of Microbiology and Immunology, University of Illinois College of Medicine, Chicago, 835 S. Wolcott (M/C 790) IL 60612-7344. star1@uic.edu

Immunoglobulin (Ig) class switch recombination (CSR) is responsible for diversification of antibody effector function during an immune response. This region-specific recombination event, between repetitive switch (S) DNA elements, is unique to B lymphocytes and is induced by activationinduced deaminase (AID). CSR is critically dependent on transcription of noncoding RNAs across S regions. However, mechanistic insight regarding this process has remained unclear. New studies indicate that long-range intrachromosomal interactions among IgH transcriptional elements organize the formation of the S/S synaptosome, as a prerequisite for CSR. This three-dimensional chromatin architecture simultaneously brings promoters and enhancers into close proximity to facilitate transcription. Here, we recount how transcription across S DNA promotes accumulation of RNA polymerase II, leading to the introduction of activating chromatin modifications and hyperaccessible chromatin that is amenable to AID activity.

Keywords: chromatin; class switch recombination; transcription; long-range interactions

Introduction

B lymphocytes undergo a set of three programmed gene alterations that diversify immunoglobulin (Ig) antigen binding capability or effector function. Antigen binding specificity is encoded by the combinatorial joining of heavy (H) chain V_H, D_H, and J_H gene segments, or of light (L) chain V_L and J_L gene segments. H- and L-chain assembly via V–D–J joining occurs in an orderly stepwise process during early B cell development in the bone marrow. Somatic hypermutation (SHM), a mutational process focused on already rearranged $V_H DJ_H$ and $V_L J_L$ region genes, leads to increased affinity of antibody binding to antigen in germinal center B cells. Class switch recombination (CSR) provides for diversification of C_H effector function while maintaining the original V(D)J rearrangement in mature B cells. IgH effector function is encoded by eight constant (C_H)

region genes (μ, δ, $\gamma 3$, $\gamma 1$, $\gamma 2b$, $\gamma 2a$, ε, and α) that span a 220 kb genomic region located near the right telomere of chromosome 12. CSR involves an intrachromosomal deletional rearrangement that focuses on a repetitive switch (S) DNA region located upstream of each C_H gene (with the exception of $C\delta$; reviewed in Refs. 1 and 2).

Germline, or noncoding, transcript (GLT) promoters associated with each of the paired S–C_H regions target CSR to specific S regions by selective transcriptional activation (reviewed in Refs. 1 and 2). GLTs are so named because they lack an open reading frame and are not translated.[3] Combinations of cytokines and B cell activators induce transcription from specific GLT promoters.[4] Gene targeting experiments indicate a critical requirement for transcription and transcriptional elements to enable CSR (reviewed in Refs. 1 and 5). Activationinduced deaminase (AID) deaminates dC residues in transcribed S regions and is essential for both CSR and SHM.[6,7] The mechanism by which AID-induced DNA lesions are repaired and culminate in SHM or CSR has been extensively reviewed elsewhere.[1,3,8–14]

---

*Current address: Cancer Vaccine Center, Department of Medical Oncology, Dana-Farber Cancer Institute, Boston, Massachusetts.

doi: 10.1111/j.1749-6632.2012.06604.x

S regions separated by as much as 150 kb can be targeted for recombination. This distance is a topological challenge to formation of the S/S synaptosome, a requirement for CSR. The mechanism by which two distant S regions targeted for recombination are brought into close proximity has been unclear. However, recent studies indicate that regulatory elements act over large genomic distances to bring separated elements into close spatial proximity with their target genes or other functional elements.[15–21] The chromatin conformation capture (3C) assay was developed to determine whether distant promoters and enhancers come into contact by long range looping or whether they serve as entry sites or nucleation points for factors that ultimately "communicate" with the gene.[22] Here, we explore recent findings regarding the influence of GLT transcription on three-dimensional chromatin architecture and its relationship to epigenetic chromatin modifications, and suggest a model of how these processes integrate to regulate the CSR reaction.

The looping out and deletion mechanism of CSR

The organization of the C_H region of the *IgH* locus is illustrated in Figure 1A. The $E\mu$ intronic enhancer is located at the 5′-end of the C_H subregion. The 3′ regulatory region, 3′$E\alpha$, contains DNase hypersensitive sites (hs) 3a, hs1, 2, hs3b, and hs4, and functions as a locus control region (LCR; reviewed in Ref. 23). Further downstream there are additional hs sites 5–7.[24] Sequestered between the two enhancers are eight C_H region genes, each paired with an S region (with the exception of $C\delta$). S region transcription is a defining feature of CSR. C_H genes are organized in transcription units consisting of a noncoding intronic (I) exon, an S region, and a C_H coding region. GLTs emanate from an I promoter located upstream of each I exon, run through the S region, and then terminate at the 3′ end of a neighboring C_H coding region (Fig. 1A).[4] Downstream S regions are selectively targeted for recombination with $S\mu$ by directed activation of the isotype-specific I exon promoters in response to combinations of antigen or mitogen, cytokines, and costimulatory signals.[2,4] Targeted deletions of I exon promoters abolish transcription and severely affect CSR frequency.[25–27]

Once transcription targets $S\mu$ and a downstream S region for recombination, CSR occurs

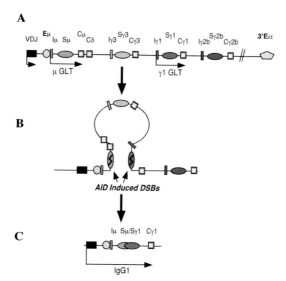

Figure 1. The looping-out and deletion model of Ig switch recombination. (A) A partial schematic map of IgH locus before CSR is shown not to scale. A productive V(D)J rearrangement has occurred, allowing expression of the γ and μ IgH chains. Intact $S\mu$ and $S\gamma 1$ are separated by approximately 70 kb. Stimulation of B cells with antigen or mitogen induces germline transcription through the $I\gamma 1$–$S\gamma 1$–$C\gamma 1$ region before recombination. (B) The $S\mu$ and $S\gamma 1$ regions are aligned causing the intervening genomic DNA to form a loop. (C) A reciprocal crossover between $S\mu$ and $S\gamma 1$ results in the formation of a new hybrid transcriptional unit containing the original VDJ exons contiguous with $C\gamma 1$ and the formation of hybrid $S\mu/S\gamma 1$ molecules.

through an intrachromosomal deletional rearrangement that results in the formation of composite $S\mu$–Sx (x = a downstream S region) junctions on the chromosome, while the intervening genomic material is looped out and excised (Fig. 1B). CSR occurs between two highly repetitive S regions.[28] Concomitant with transcription, AID initiates S region–specific double strand breaks (DSBs)[29,30] that are processed through a cascade of events mediated by nonhomologous end joining (NHEJ).[3] CSR occurs anywhere between the donor $S\mu$ and an acceptor S region. Each CSR event produces a new composite S/S junction, which as a population is heterogeneous with respect to the position of the recombination crossover points (Fig. 1C). However, it has been unclear how distantly located S region-specific DSBs, separated by as much as 150 kb, are recruited to partner in a CSR reaction that leads to intrachromosomal deletion.

A

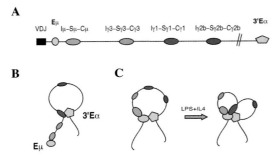

B

C

Figure 2. The IgH locus is configured in chromatin loops in B cells. (A) A schematic is shown of the linear IgH locus with the VDJ exons (black box), intronic Eμ and 3′Eα enhancers (yellow symbols), C$_H$ regions (gray boxes), S regions (ovals), and I exons (rectangles) indicated. This linear configuration is typical for the IgH locus in splenic T cells.[36] (B) Association of the γ1 locus with the 3′Eα will promote γ1 GLT expression but not S/S synapsis. (C) In resting mature splenic B cells the IgH locus loop is tethered by interactions between Eμ and 3′Eα. Following treatment of resting B cells with LPS and IL-4 for 40 h, Eμ and 3′Eα remain in close proximity, and the γ1 locus has been recruited to the 3′Eα regulatory region.[36] Sγ1 is now in close proximity to Sμ.

The mechanism of CSR is tied to the three-dimensional structure of the IgH locus

Long-range chromatin interactions over large genomic distances bring widely separated elements into close spatial proximity to their target genes. The mouse β-globin genes and their LCR, located more than 50 kb apart, engage in higher-order loop structures during transcription, and these interactions are tightly correlated with gene-specific expression.[16,17,31] Looped chromatin structures involved in the regulation of gene expression have also been detected in association with imprinted genes,[19,32,33] cytokine clusters,[18] estrogen receptor,[15] the major histocompatibility complex (MHC),[21] α globin[20] loci, and between chromatin boundary elements.[34] In the IgH locus, inducible transcription from downstream GLT promoters requires the 3′Eα LCR, as targeted deletion of hs3b,4 within 3′Eα leads to loss of GLT expression.[2,35] The direct interaction between GLT promoters and distant 3′Eα may be mediated by chromatin loop formation. But while spatial proximity between the various downstream GLT promoters with 3′Eα may facilitate transcription, it does not support S/S synapsis because the S regions would not be juxtaposed with Sμ (Fig. 2A and B). Our published 3C studies indicated that in mature resting B cells, the Eμ and 3′Eα en-

hancers are in close spatial proximity and form a chromatin loop (Fig. 2C).[36] B cell activation leads to cytokine-dependent recruitment of the GLT promoters to the Eμ:3′Eα complex and enables transcription of S regions targeted for CSR (Fig. 2C).[36] Our new data indicate that the GLT promoter (and not the S regions) directly interacts with the 3′Eα element (unpublished observations). Before CSR, this looped structure facilitates S–S synapsis because Sμ is proximal to Eμ and a downstream S region is corecruited with the targeted GLT promoter to the Eμ:3′Eα complex. We and others conclude the IgH locus assumes a three-dimensional structure that simultaneously supports GLT promoter interactions with the 3′Eα regulatory regions to facilitate transcription of the targeted S regions and juxtaposes the S regions targeted for recombination.[36–38]

S regions are specialized targets of CSR

S regions are G/C rich and fall into two general groups: sequences incorporating prevalent pentameric repeats into larger tandem repeats, such as found in Sμ, Sε, and Sα, and a 49 bp repeat that is characteristic of Sγ3, Sγ1, Sγ2b, and Sγ2a.[28] S/S junctions are dispersed throughout S regions and occasionally fall in areas flanking the S regions.[39,40] CSR junctions are distinguished by their lack of consensus sequences or significant homology from site-specific or homologous recombination. S region length influences CSR frequency both *in vivo* and *in vitro*.[39,41] Deletion of S regions or their replacement with non-S region sequences by gene targeting methods reduces CSR, indicating that S regions are the specialized targets in this recombination event.[41–44]

An intriguing question is, how are S regions with their divergent sequences recognized by AID and the CSR machinery? The degeneracy of the S region repeats and the absence of an identifiable recombination motif have led to models in which higher-order structures, rather than primary sequence, provides the recognition code for the CSR machinery.[2,45] S regions are replete with palindromic sequences that have the potential to form stem–loop structures that are proposed to function as recognition substrates.[45–47] Murine and human S regions are G-rich on the nontemplate strand, which contributes to R-loop formation[48,49] and G4 tetraplexes (see Refs. 50 and 94). Transcribed S regions contain R-loops of RNA:DNA hybrids *in vitro* and *in vivo*, which can provide ssDNA stretches as a substrate for AID deamination.[48,49] Targeted inversion of the Sγ1 region, resulting in the G-rich strand being nontemplate, leads to the loss of R-loop formation and a significant reduction of CSR activity, which indicates the importance of R-loops for CSR efficiency.[44]

Transcription and asymmetric AID attack

Early studies suggested that AID may function as an RNA deaminase, based on its similarities to APOBEC1, the catalytic component of a tissue-specific RNA editing complex (Refs. 51, 52, and 94). Identification of a mutational hotspot for AID in SHM indicated a mechanism by which AID functions directly on DNA.[53] This observation prompted Neuberger and coworkers to propose a DNA deamination model for AID action in which AID initiates SHM and CSR by converting deoxycytidine (dC) to deoxyuracil (dU) that can then be processed by one of several mechanisms (Fig. 3; reviewed in Refs. 1, 3, 8, and 14). The DNA deamination model gained substantial support from the direct demonstration that AID-dependent dU residues are detected in the 5'Sμ region,[54] and that uracil DNA glycosylase (UNG) is required for formation of DSB in S regions to enable CSR.[29]

During CSR, AID-induced S region mutations begin 150 bp downstream of the I exon transcription start site (TSS)[55] (Fig. 4). The 3' boundary for CSR mutations is located downstream of S regions at a distance of up to 10 kb from the TSS[55] (Fig. 4). Studies indicate that ectopically expressed AID binds with RNA polymerase II (RNAP II) in coimmunoprecipitation assays[56] and interacts with the transcription apparatus *in vitro*,[57] suggesting that AID targeting is linked with transcription. Surprisingly, AID attack is asymmetrically focused within the I–S–C$_H$ region. S regions are substrates for AID activity, whereas C$_H$ regions within the same transcription unit are protected,[29,30,55] raising the question of how this occurs.

Chromatin accessibility to AID attack

Eukaryotic DNA is wrapped around histone octamers and organized into higher-order chromatin fibers that regulate access of *trans*-acting factors to DNA.[58–60] Studies suggest that at least two distinct mechanisms are used to achieve efficient transcription through chromatin. One relies on a pathway based on nucleosome loss, and the second involves histone acetylation with

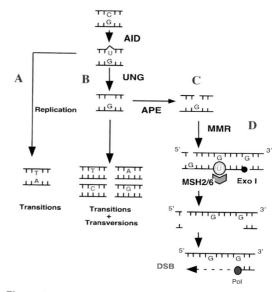

Figure 3. AID deamination model and a role for mismatch repair (MMR) in CSR. AID deaminates dC to dU. (A) The U:G mismatch can be replicated over to produce T:A, a transition mutation. (B) Using the base excision repair (BER) pathway, dU bases can be excised by a uracil DNA glycosylase (UNG), leaving an abasic site that can then be replicated over by an error prone polymerase to produce both transition and transversion mutations. (C) Abasic sites can also be recognized by AP endonucleases (APE) to form ssDNA nicks or double strand breaks (DSBs) when the abasic sites are closely spaced and on complementary strands. (D) Other U:G mismatches could be substrates for mismatch repair (MMR) MSH2/MSH6 binding, which in turn recruit PMS2-MLH1 (not shown), and PMS2 nicks the DNA. Exonuclease 1 (Exo I) binds to MSH2 and MLH1, and using the PMS2-induced nick, excises toward the U:G mismatch, producing a DSB with a 5' overhang. An error-prone DNA polymerase can fill in the 5' overhang and introduces mutations at A:T bp; 3' overhangs can be excised by Ercc1-XPF. Short overhangs can be used during end joining.

little or no loss of histones (reviewed in Ref. 61). Our studies indicate that the second pathway, involving histone modification, occurs during GLT expression during CSR[62,63] (described below). The RNAP II transcription cycle occurs in distinct steps, including RNAP II recruitment to the promoter, formation of the preinitiation complex, promoter clearance, processive elongation, and transcription termination.[61] When RNAP II clears the promoter and transits to the promoter proximal pause site, it becomes hyperphosphorylated at Ser5 (p-ser5) on its C-terminal domain (CTD). RNAP II then enters into productive elongation, loses p-ser5, and becomes enriched for p-ser2

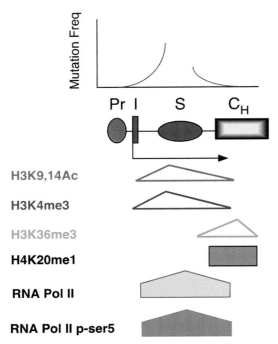

Figure 4. Summary of mutation frequency and differential histone modifications in S and C_H regions of activated B cells. A schematic diagram depicts a generic I–S–C_H region downstream of the GLT promoter (Pr). Above the diagram, a summary of mutation frequencies 5′ and 3′ of the S region for Ung[−/−] Msh2[−/−] B cells is shown.[56] A similar distribution of mutations are found for this region in Ung[−/−] B cells.[82] Below the I–S–C_H schematic, a summary of histone modifications and RNA pol II binding is shown. Wild-type B cells were stimulated with LPS or LPS + IL-4 for 48 h and then analyzed for mutations by chromatin immunoprecipitation (ChIP) assays using antisera against H3K9,14Ac, H3K4me3, H3K36me3, H4K20me1, pol II RNA, and pol II p-ser5. Data were amalgamated from published studies.[74,75]

in the CTD.[61] An emerging paradigm links transcription regulation with the phosphorylation status of the RNAP II CTD and the recruitment of histone-modifying enzymes, which in turn introduce histone marks that alter the status of chromatin accessibility.[59,64] This paradigm is relevant to understanding asymmetric AID targeting away from C_H regions and toward S regions.

Transcription activity is strikingly correlated with significant histone acetylation (Ac) at promoters.[65,66] Genome-wide studies reveal that promoter-proximal sites in transcriptionally active genes are enriched for trimethyl histone H3 lysine 4 (H3K4me3), hyperacetylated (Ac) H3K9, and RNAP II p-ser5, whereas downstream coding regions are enriched in H3K36me3 and elongation-

associated RNAP II p-ser2.[66–69] Histone methylation can act as a tag for effector proteins containing methyl-binding domains, including chromodomains, tudor domains, and plant homeodomain (PHD) finger domains.[70] The NuA3 histone acetyltransferase (HAT) complex that coordinates transcription activation with histone Ac is directly bound by H3K4me3 (reviewed in Ref. 71). Histone Ac, which changes the net charge of nucleosomes and alters chromatin fiber–folding properties, increases DNA accessibility.[72] Histone Ac may also create binding surfaces for factor–histone contacts that then facilitate recruitment of transcription regulators.[59] Rpd3S, a histone deactylase (HDAC) complex, interacts with H3K36me3, which then functions to reduce histone Ac and thereby repress transcription initiation in downstream coding regions.[73,74] The bifurcation of H3K4me3 and H3K36me3 modifications within a transcription unit serve to recruit HATs and HDACs, respectively, which in turn modulate chromatin accessibility.

Consistent with the observation that S regions, but not C_H regions, are modified by AID, markers of accessible chromatin are associated with S regions but not with C_H regions. *In vivo*, I–S regions undergoing active transcription are nuclease hypersensitive, whereas C_H regions are comparatively inaccessible.[62,63] Accordingly, antisense RNA transcripts are found selectively in S regions but not in C_H regions, indicating reciprocal areas of accessible or repressed chromatin, respectively.[75] Transcriptionally active I–S regions accumulate H3K4me3 and histone Ac modifications, while C_H regions become enriched for the repressive countermarks H3K36me3 and H4K20me1[56,62,63,76] (Fig. 4). Treatment with tricostatin A, an inhibitor of HDACs, increases H3Ac at S regions concomitant with greater chromatin accessibility and higher frequency CSR, demonstrating a direct link between histone Ac and switching frequency.[63] Furthermore, PTIP (PAX interaction with transcription activation domain protein)-deficiency displays reduced H3K4me3 and histone Ac marks and impaired transcription initiation at a subset of downstream S regions.[77] PTIP is a component of several histone methyl transferases (HMTases), including mixed lineage leukemia 3 (MLL3)–MLL4 complex, that interact with RNAPII p-ser5.[77,78] PTIP also indirectly associates with p300/CBP HAT, suggesting that PTIP regulates both histone Ac and Me.[79]

Interestingly, long-range looping interactions between GLT promoters and the 3′Eα enhancer are impaired in PTIP-deficient B cells, indicative of deficient GLT transcription initiation.[80] Collectively, these observations support the conclusion that histone Ac status and chromatin accessibility in S regions are determined by H3K4me3 modifications linked to transcription.

S regions, R-loops, and AID targeting

S regions span 1–10 kb beyond I exons and TSSs,[28] and when transcribed, S regions are enriched along their *entire* lengths with H3K4me3-, H3Ac-, and H4Ac- activating histone modifications.[62,63] This contrasts with genome-wide analyses in which these marks are largely relegated to promoter-proximal locations.[66,67] What directs the spread of these modifications long distances from the GLT TSSs? This unusual distribution of H3K4me3 in S regions is linked to stalled RNAP II p-ser5 in S region DNA. Chromatin immunoprecipitation (ChIP) studies showed accumulation of RNAP II p-ser5 throughout the Sμ and Sγ3 regions, indicating that RNAP II has stalled at multiple sites.[63] In contrast, RNAP II p-ser5 occupancy and H3K4me3 marks remained promoter proximal in Sμ-deleted B cells, indicating that it is the S region sequence that stalls RNAP II, leading to the unusual H3K4me3 distribution.[63,81] S regions are highly G/C rich and when transcribed develop R-loops over long stretches that enhance CSR efficiency *in vivo* (see Refs. 48, 49, and 93). It is likely that R-loops that demonstrably impede transcription elongation[82,83] are responsible for RNAP II accumulation detected in S regions. Further investigation is required to ascribe a causal relationship between R-loop formation and RNAP II accumulation in S regions.

Is there a connection between RNAP II stalling and focused AID activity on S regions? Recent evidence suggests that AID interacts with the transcription elongation factor Spt5. Spt5 was identified in an shRNA screen as a critical regulator of CSR.[84] Upon clearance from the promoter, RNAP II pausing occurs 25–50 nucleotides downstream of the TSS and is mediated by DSIF (5,6-dichloro-1-β-ᴅ-ribofuranosylbenzimidazole (DRB) sensitivity-inducing factor), comprising Spt4, Spt5, and negative elongation factor (NEF).[85] Spt5 interacts with AID and mediates its association with RNAP II.[84] ChIP-seq experiments indicate overlapping positions of Spt5 binding with RNAP II stalling[84,86] and between sites of Spt5 and AID binding genome-wide.[84] Spt5 also interacts with several cotranscriptional factors, including splicing factors,[87] capping enzyme,[88] and the RNA exosome,[89] that have been functionally associated with CSR.[90–92] Targeted deletion of the GLT splice donor element dramatically impaired CSR.[25] These observations highlight the highly integrated aspect of transcription elongation and RNA processing. Together, these findings suggest that AID is targeted to sites of high-occupancy RNAP II by means of its association with Spt5. The unique character of S regions, which are prone to forming R-loops, in turn impedes RNAP II elongation. It now seems that Spt5, a cofactor of stalled RNAP II, associates with AID and may function to target it to specific loci. Because C_H regions lack R-loops, stalled RNAP II does not accumulate, and Spt5 does not focus AID to these regions.

Summary

Transcription is intrinsic to the mechanism of CSR. The unique structure of the S regions appears to be mechanistically linked to generating open chromatin. These advances highlight the nexus between transcription, long-range looping interactions with chromatin remodeling, and histone modifications, which permit initiation of CSR by AID. Emerging new technologies that allow visualization of higher-order chromatin structures are likely to provide new insights into transcription regulation by regulatory elements functioning at great distances from their target genes. Ig CSR is an excellent model to study the intersection of long-range chromatin interactions, transcription, and recombination.

Acknowledgment

This work was supported in part by the National Institutes of Health (AI052400 to A.L.K.).

Conflicts of interest

The authors declare that they have no competing financial interests or conflicts of interest.

References

1. Stavnezer, J., J.E. Guikema & C.E. Schrader. 2008. Mechanism and regulation of class switch recombination. *Annu. Rev. Immunol.* **26:** 261–292.
2. Manis, J.P., M. Tian & F.W. Alt. 2002. Mechanism and control of class-switch recombination. *Trends Immunol.* **23:** 31–39.

3. Chaudhuri, J., U. Basu, A. Zarrin, *et al.* 2007. Evolution of the immunoglobulin heavy chain class switch recombination mechanism. *Adv. Immunol.* **94:** 157–214.

4. Stavnezer, J. 2000. Molecular processes that regulate class switching. *Curr. Top Microbiol. Immunol.* **245:** 127–168.

5. Perlot, T. & F.W. Alt. 2008. Cis-regulatory elements and epigenetic changes control genomic rearrangements of the IgH locus. *Adv. Immunol.* **99:** 1–32.

6. Muramatsu, M., K. Kinoshita, S. Fagarasan, *et al.* 2000. Class switch recombination and hypermutation require activation-induced cytidine deaminase (AID), a potential RNA editing enzyme. *Cell* **102:** 553–563.

7. Revy, P., T. Muto, Y. Levy, *et al.* 2000. Activation-induced cytidine deaminase (AID) deficiency causes the autosomal recessive form of the Hyper-IgM syndrome (HIGM2). *Cell* **102:** 565–575.

8. Peled, J.U., F.L. Kuang, M.D. Iglesias-Ussel, *et al.* 2008. The biochemistry of somatic hypermutation. *Annu. Rev. Immunol.* **26:** 481–511.

9. Neuberger, M.S., J.M. Di Noia, R.C. Beale, *et al.* 2005. Somatic hypermutation at A.T pairs: polymerase error versus dUTP incorporation. *Nat. Rev. Immunol.* **5:** 171–178.

10. Saribasak, H., D. Rajagopal, R.W. Maul & P.J. Gearhart. 2009. Hijacked DNA repair proteins and unchained DNA polymerases. *Philos. Trans. R. Soc. Lond. B Biol. Sci.* **364:** 605–611.

11. Teng, G. & F.N. Papavasiliou. 2007. Immunoglobulin somatic hypermutation. *Annu. Rev. Genet.* **41:** 107–120.

12. Kenter, A.L. 2005. Class switch recombination: an emerging mechanism. *Curr. Top. Microbiol. Immunol.* **290:** 171–199.

13. Stavnezer, J. 2011. Complex regulation and function of activation-induced cytidine deaminase. *Trends Immunol.* **32:** 194–201.

14. Di Noia, J.M. & M.S. Neuberger. 2007. Molecular mechanisms of antibody somatic hypermutation. *Annu. Rev. Biochem.* **76:** 1–22.

15. Fullwood, M.J., M.H. Liu, Y.F. Pan, *et al.* 2009. An oestrogen-receptor-alpha-bound human chromatin interactome. *Nature* **462:** 58–64.

16. Vakoc, C.R., D.L. Letting, N. Gheldof, *et al.* 2005. Proximity among distant regulatory elements at the beta-globin locus requires GATA-1 and FOG-1. *Mol. Cell* **17:** 453–462.

17. Tolhuis, B., R.J. Palstra, E. Splinter, *et al.* 2002. Looping and interaction between hypersensitive sites in the active beta-globin locus. *Mol. Cell* **10:** 1453–1465.

18. Spilianakis, C.G. & R.A. Flavell. 2004. Long-range intrachromosomal interactions in the T helper type 2 cytokine locus. *Nat. Immunol.* **5:** 1017–1027.

19. Kurukuti, S., V.K. Tiwari, G. Tavoosidana, *et al.* 2006. CTCF binding at the H19 imprinting control region mediates maternally inherited higher-order chromatin conformation to restrict enhancer access to Igf2. *Proc. Natl. Acad. Sci. USA* **103:** 10684–10689.

20. Zhou, G.L., L. Xin, W. Song, *et al.* 2006. Active chromatin hub of the mouse alpha-globin locus forms in a transcription factory of clustered housekeeping genes. *Mol. Cell Biol.* **26:** 5096–5105.

21. Majumder, P., J.A. Gomez, B.P. Chadwick & J.M. Boss. 2008. The insulator factor CTCF controls MHC class II gene expression and is required for the formation of long-distance chromatin interactions. *J. Exp. Med.* **205:** 785–798.

22. Dekker, J., K. Rippe, M. Dekker & N. Kleckner. 2002. Capturing chromosome conformation. *Science* **295:** 1306–1311.

23. Khamlichi, A.A., E. Pinaud, C. Decourt, *et al.* 2000. The 3′ IgH regulatory region: a complex structure in a search for a function. *Adv. Immunol.* **75:** 317–345.

24. Garrett, F.E., A.V. Emelyanov, M.A. Sepulveda, *et al.* 2005. Chromatin architecture near a potential 3′ end of the igh locus involves modular regulation of histone modifications during B-Cell development and in vivo occupancy at CTCF sites. *Mol. Cell Biol.* **25:** 1511–1525.

25. Jung, S., K. Rajewsky & A. Radbruch. 1993. Shutdown of class switch recombination by deletion of a switch region control element. *Science* **259:** 984–987.

26. Zhang, J., A. Bottaro, S. Li, *et al.* 1993. A selective defect in IgG2b switching as a result of targeted mutation of the Ig2b promoter and exon. *EMBO J.* **12:** 3529–3537.

27. Seidl, K.J., J.P. Manis, A. Bottaro, *et al.* 1999. Position-dependent inhibition of class-switch recombination by PGK-neor cassettes inserted into the immunoglobulin heavy chain constant region locus. *Proc. Natl. Acad. Sci. USA* **96:** 3000–3005.

28. Gritzmacher, C.A. 1989. Molecular aspects of heavy-chain class switching. *Crit. Rev. Immunol.* **9:** 173–200.

29. Schrader, C.E., E.K. Linehan, S.N. Mochegova, *et al.* 2005. Inducible DNA breaks in Ig S regions are dependent on AID and UNG. *J. Exp. Med.* **202:** 561–568.

30. Wuerffel, R.A., J. Du, R.J. Thompson & A.L. Kenter. 1997. Ig Sgamma3 DNA-specifc double strand breaks are induced in mitogen-activated B cells and are implicated in switch recombination. *J. Immunol.* **159:** 4139–4144.

31. Carter, D., L. Chakalova, C.S. Osborne, *et al.* 2002. Long-range chromatin regulatory interactions in vivo. *Nat. Genet.* **32:** 623–626.

32. Murrell, A., S. Heeson & W. Reik. 2004. Interaction between differentially methylated regions partitions the imprinted genes Igf2 and H19 into parent-specific chromatin loops. *Nat. Genet.* **36:** 889–893.

33. Horike, S., S. Cai, M. Miyano, *et al.* 2005. Loss of silent-chromatin looping and impaired imprinting of DLX5 in Rett syndrome. *Nat. Genet.* **37:** 31–40.

34. Blanton, J., M. Gaszner & P. Schedl. 2003. Protein: protein interactions and the pairing of boundary elements in vivo. *Genes Dev.* **17:** 664–675.

35. Pinaud, E., A.A. Khamlichi, C. Le Morvan, *et al.* 2001. Localization of the 3′ IgH locus elements that effect long-distance regulation of class switch recombination. *Immunity* **15:** 187–199.

36. Wuerffel, R., L. Wang, F. Grigera, *et al.* 2007. S-S synapsis during class switch recombination is promoted by distantly located transcriptional elements and activation-induced deaminase. *Immunity* **27:** 711–722.

37. Ju, Z., S.A. Volpi, R. Hassan, *et al.* 2007. Evidence for physical interaction between the immunoglobulin heavy chain variable region and the 3′ regulatory region. *J. Biol. Chem.* **282:** 35169–35178.

38. Sellars, M., B. Reina-San-Martin, P. Kastner & S. Chan. 2009. Ikaros controls isotype selection during

immunoglobulin class switch recombination. *J. Exp. Med.* **206**: 1073–1087.

39. Kenter, A.L., R. Wuerffel, C. Dominguez, *et al.* 2004. Mapping of a functional recombination motif that defines isotype specificity for mu–>gamma3 switch recombination implicates NF-kappaB p50 as the isotype-specific switching factor. *J. Exp. Med.* **199**: 617–627.

40. Schrader, C.E., S.P. Bradley, J. Vardo, *et al.* 2003. Mutations occur in the Ig Smu region but rarely in Sgamma regions prior to class switch recombination. *EMBO J.* **22**: 5893–5903.

41. Zarrin, A.A., M. Tian, J. Wang, *et al.* 2005. Influence of switch region length on immunoglobulin class switch recombination. *Proc. Natl. Acad. Sci. USA* **102**: 2466–2470.

42. Khamlichi, A.A., F. Glaudet, Z. Oruc, *et al.* 2004. Immunoglobulin class-switch recombination in mice devoid of any Smu tandem repeat. *Blood* **103**: 3828–3836.

43. Luby, T.M., C.E. Schrader, J. Stavnezer & E. Selsing. 2001. The mu switch region tandem repeats are important, but not required, for antibody class switch recombination. *J. Exp. Med.* **193**: 159–168.

44. Shinkura, R., M. Tian, M. Smith, *et al.* 2003. The influence of transcriptional orientation on endogenous switch region function. *Nat. Immunol.* **4**: 435–441.

45. Honjo, T., K. Kinoshita & M. Muramatsu. 2002. Molecular mechanism of class switch recombination: linkage with somatic hypermutation. *Annu. Rev. Immunol.* **20**: 165–196.

46. Tashiro, J., K. Kinoshita & T. Honjo. 2001. Palindromic but not G-rich sequences are targets of class switch recombination. *Int. Immunol.* **13**: 495–505.

47. Mussmann, R., M. Courtet, J. Schwager, L. Du Pasquier. 1997. Microsites for immunoglobulin switch recombination breakpoints from Xenopus to mammals. *Eur. J. Immunol.* **27**: 2610–2619.

48. Yu, K., F. Chedin, C.L. Hsieh, *et al.* 2003. R-loops at immunoglobulin class switch regions in the chromosomes of stimulated B cells. *Nat. Immunol.* **4**: 442–451.

49. Huang, F.T., K. Yu, B.B. Balter, *et al.* 2007. Sequence dependence of chromosomal R-loops at the immunoglobulin heavy-chain Smu class switch region. *Mol. Cell Biol.* **27**: 5921–5932.

50. Duquette, M.L., P. Handa, J.A. Vincent, *et al.* 2004. Intracellular transcription of G-rich DNAs induces formation of G-loops, novel structures containing G4 DNA. *Genes. Dev.* **18**: 1618–1629.

51. Blanc, V., S. Kennedy & N.O. Davidson. 2003. A novel nuclear localization signal in the auxiliary domain of apobec-1 complementation factor regulates nucleocytoplasmic import and shuttling. *J. Biol. Chem.* **278**: 41198–41204.

52. Muramatsu, M., V.S. Sankaranand, S. Anant, *et al.* 1999. Specific expression of activation-induced cytidine deaminase (AID), a novel member of the RNA-editing deaminase family in germinal center B cells. *J. Biol. Chem.* **274**: 18470–18476.

53. Rada, C., M.R. Ehrenstein, M.S. Neuberger & C. Milstein. 1998. Hot spot focusing of somatic hypermutation in MSH2-deficient mice suggests two stages of mutational targeting. *Immunity* **9**: 135–141.

54. Maul, R.W. & P.J. Gearhart. 2010. AID and somatic hypermutation. *Adv. Immunol.* **105**: 159–191.

55. Xue, K., C. Rada & M.S. Neuberger. 2006. The in vivo pattern of AID targeting to immunoglobulin switch regions deduced from mutation spectra in msh2–/– ung–/– mice. *J. Exp. Med.* **203**: 2085–2094.

56. Nambu, Y., M. Sugai, H. Gonda, *et al.* 2003. Transcription-coupled events associating with immunoglobulin switch region chromatin. *Science* **302**: 2137–2140.

57. Besmer, E., E. Market & F.N. Papavasiliou. 2006. The transcription elongation complex directs activation-induced cytidine deaminase-mediated DNA deamination. *Mol. Cell Biol.* **26**: 4378–4385.

58. Woodcock, C.L. & S. Dimitrov. 2001. Higher-order structure of chromatin and chromosomes. *Curr. Opin. Genet. Dev.* **11**: 130–135.

59. Li, B., M. Carey & J.L. Workman. 2007. The role of chromatin during transcription. *Cell* **128**: 707–719.

60. Suganuma, T. & J.L. Workman. 2011. Signals and combinatorial functions of histone modifications. *Annu. Rev. Biochem.* **80**: 473–499.

61. Selth, L.A., S. Sigurdsson & J.Q. Svejstrup. 2010. Transcript elongation by RNA polymerase II. *Annu. Rev. Biochem.* **79**: 271–293.

62. Wang, L., N. Whang, R. Wuerffel & A.L. Kenter. 2006. AID-dependent histone acetylation is detected in immunoglobulin S regions. *J. Exp. Med.* **203**: 215–226.

63. Wang, L., R. Wuerffel, S. Feldman, *et al.* 2009. S region sequence, RNA polymerase II, and histone modifications create chromatin accessibility during class switch recombination. *J. Exp. Med.* **206**: 1817–1830.

64. Saunders, A., L.J. Core & J.T. Lis. 2006. Breaking barriers to transcription elongation. *Nat. Rev. Mol. Cell Biol.* **7**: 557–567.

65. Workman, J.L. & R.E. Kingston. 1998. Alteration of nucleosome structure as a mechanism of transcriptional regulation. *Annu. Rev. Biochem.* **67**: 545–579.

66. Pokholok, D.K., C.T. Harbison, S. Levine, *et al.* 2005. Genome-wide map of nucleosome acetylation and methylation in yeast. *Cell* **122**: 517–527.

67. Bernstein, B.E., M. Kamal, K. Lindblad-Toh, *et al.* 2005. Genomic maps and comparative analysis of histone modifications in human and mouse. *Cell* **120**: 169–181.

68. Barski, A., S. Cuddapah, K. Cui, *et al.* 2007. High-resolution profiling of histone methylations in the human genome. *Cell* **129**: 823–837.

69. Guenther, M.G., S.S. Levine, L.A. Boyer, *et al.* 2007. A chromatin landmark and transcription initiation at most promoters in human cells. *Cell* **130**: 77–88.

70. Daniel, J.A., M.G. Pray-Grant & P.A. Grant. 2005. Effector proteins for methylated histones: an expanding family. *Cell Cycle* **4**: 919–926.

71. Berger, S.L. 2007. The complex language of chromatin regulation during transcription. *Nature* **447**: 407–412.

72. Shahbazian, M.D. & M. Grunstein. 2007. Functions of site-specific histone acetylation and deacetylation. *Annu. Rev. Biochem.* **76**: 75–100.

73. Carrozza, M.J., B. Li, L. Florens, *et al.* 2005. Histone H3 methylation by Set2 directs deacetylation of coding regions by Rpd3S to suppress spurious intragenic transcription. *Cell* **123**: 581–592.

74. Lieb, J.D. & N.D. Clarke. 2005. Control of transcription through intragenic patterns of nucleosome composition. *Cell* **123:** 1187–1190.

75. Perlot, T., G. Li & F.W. Alt. 2008. Antisense transcripts from immunoglobulin heavy-chain locus V(D)J and switch regions. *Proc. Natl. Acad. Sci. USA* **105:** 3843–3848.

76. Li, Z., Z. Luo & M.D. Scharff. 2004. Differential regulation of histone acetylation and generation of mutations in switch regions is associated with Ig class switching. *Proc. Natl. Acad. Sci. USA* **101:** 15428–15433.

77. Daniel, J.A., M.A. Santos, Z. Wang, *et al.* 2010. PTIP promotes chromatin changes critical for immunoglobulin class switch recombination. *Science* **329:** 917–923.

78. Munoz, I.M. & J. Rouse. 2009. Control of histone methylation and genome stability by PTIP. *EMBO Rep.* **10:** 239–245.

79. Hoffmeister, A., A. Ropolo, S. Vasseur, *et al.* 2002. The HMG-I/Y-related protein p8 binds to p300 and Pax2 trans-activation domain-interacting protein to regulate the trans-activation activity of the Pax2A and Pax2B transcription factors on the glucagon gene promoter. *J. Biol. Chem.* **277:** 22314–22319.

80. Schwab, K.R., S.R. Patel & G.R. Dressler. 2011. Role of PTIP in class switch recombination and long-range chromatin interactions at the immunoglobulin heavy chain locus. *Mol. Cell Biol.* **31:** 1503–1511.

81. Rajagopal, D., R.W. Maul, A. Ghosh, *et al.* 2009. Immunoglobulin switch mu sequence causes RNA polymerase II accumulation and reduces dA hypermutation. *J. Exp. Med.* **206:** 1237–1244.

82. Huertas, P. & A. Aguilera. 2003. Cotranscriptionally formed DNA:RNA hybrids mediate transcription elongation impairment and transcription-associated recombination. *Mol. Cell* **12:** 711–721.

83. Tous, C. & A. Aguilera. 2007. Impairment of transcription elongation by R-loops in vitro. *Biochem. Biophys. Res. Commun.* **360:** 428–432.

84. Pavri, R. & M.C. Nussenzweig. 2011. AID targeting in antibody diversity. *Adv. Immunol.* **110:** 1–26.

85. Fuda, N.J., M.B. Ardehali & J.T. Lis. 2009. Defining mechanisms that regulate RNA polymerase II transcription in vivo. *Nature* **461:** 186–192.

86. Rahl, P.B., C.Y. Lin, A.C. Seila, *et al.* 2010. c-Myc regulates transcriptional pause release. *Cell* **141:** 432–445.

87. Pei, Y. & S. Shuman. 2002. Interactions between fission yeast mRNA capping enzymes and elongation factor Spt5. *J. Biol. Chem.* **277:** 19639–19648.

88. Wen, Y. & A.J. Shatkin. 1999. Transcription elongation factor hSPT5 stimulates mRNA capping. *Genes. Dev.* **13:** 1774–1779.

89. Andrulis, E.D., J. Werner, A. Nazarian, *et al.* 2002. The RNA processing exosome is linked to elongating RNA polymerase II in Drosophila. *Nature* **420:** 837–841.

90. Ganesh, K., S. Adam, B. Taylor, *et al.* 2011. CTNNBL1 is a novel nuclear localization sequence-binding protein that recognizes RNA-splicing factors CDC5L and Prp31. *J. Biol. Chem.* **286:** 17091–17102.

91. Conticello, S.G., K. Ganesh, K. Xue, *et al.* 2008. Interaction between antibody-diversification enzyme AID and spliceosome-associated factor CTNNBL1. *Mol. Cell* **31:** 474–484.

92. Nowak, U., A.J. Matthews, S. Zheng & J. Chaudhuri. 2011. The splicing regulator PTBP2 interacts with the cytidine deaminase AID and promotes binding of AID to switch-region DNA. *Nat. Immunol.* **12:** 160–166.

93. Maizels, N. 2012. G4 motifs in human genes. *Ann. N.Y. Acad. Sci.* **1267:** 53–60. This volume.

94. Conticello, S.G. 2012. Creative deaminases, self-inflicted damage, and genome evolution. *Ann. N.Y. Acad. Sci.* **1267:** 79–85. This volume.

Ann. N.Y. Acad. Sci. ISSN 0077-8923

ANNALS OF THE NEW YORK ACADEMY OF SCIENCES
Issue: *Effects of Genome Structure and Sequence on Variation and Evolution*

Preaching about the converted: how meiotic gene conversion influences genomic diversity

Francesca Cole,[1] Scott Keeney,[2,3] and Maria Jasin[1]

[1]Developmental Biology Program, [2]Molecular Biology Program, and [3]Howard Hughes Medical Institute, Memorial Sloan-Kettering Cancer Center, 1275 York Avenue, New York, New York

Address for correspondence: Francesca Cole, Developmental Biology Program, Box 109, Memorial Sloan-Kettering Cancer Center, 1275 York Avenue, New York, NY 10065. ColeF@mskcc.org or FrancescaColePhD@gmail.com

Meiotic crossover (CO) recombination involves a reciprocal exchange between homologous chromosomes. COs are often associated with gene conversion at the exchange site where genetic information is unidirectionally transferred from one chromosome to the other. COs and independent assortment of homologous chromosomes contribute significantly to the promotion of genomic diversity. What has not been appreciated is the contribution of another product of meiotic recombination, noncrossovers (NCOs), which result in gene conversion without exchange of flanking markers. Here, we review our comprehensive analysis of recombination at a highly polymorphic mouse hotspot. We found that NCOs make up ∼90% of recombination events. Preferential recombination initiation on one chromosome allowed us to estimate the contribution of CO and NCO gene conversion to transmission distortion, a deviation from Mendelian inheritance in the population. While NCO gene conversion tracts are shorter, and thus have a more punctate effect, their higher frequency translates into an approximately two-fold greater contribution than COs to gene conversion–based allelic shuffling and transmission distortion. We discuss the potential impact of mammalian NCO characteristics on evolution and genomic diversity.

Keywords: noncrossover; gene conversion; recombination; meiosis; hotspot

Introduction

Meiosis is a specialized cell division program in which a diploid precursor cell undergoes one round of DNA replication followed by two consecutive rounds of DNA segregation and cellular division to generate haploid gametes for sexual reproduction. The first cellular division cycle, meiosis I, is a reductional division in that homologous chromosomes (homologs) segregate to daughter cells, resulting in half the chromosomal complement (e.g., in humans, 23 homolog pairs are reduced to 23 chromosomes), while during the second cellular division cycle, meiosis II, sister chromatids segregate.[1] It can be argued that the primary mandate of meiosis I is to induce homologs to find each other, stably pair, and accurately segregate[2] (see also Kauppi *et al.*[3]). Failures in this process lead to gamete aneuploidy, which is the leading cause of developmental disability and spontaneous miscarriage in humans.[4] However, the DNA interactions ensuring the meiotic reductional division also have critically important consequences for evolution and genomic diversity.

Meiotic recombination: damaging the genome in order to propagate it

Meiosis has coopted ancient DNA repair mechanisms that predate sexual reproduction to provoke homologs to stably pair.[5] To make use of these mechanisms, meiotic cells induce programmed DNA double-strand breaks (DSBs) at hotspots throughout the genome by expressing the SPO11 transesterase (Fig. 1).[6] The location of meiotic hotspots is variable and will be discussed further below. DSB resection generates $3'$ single-stranded tails that are bound by strand invasion proteins to catalyze D-loop formation with an intact homologous duplex that serves as a template for DNA repair synthesis. The homolog is the preferred repair template in meiosis rather than the identical sister chromatid,

doi: 10.1111/j.1749-6632.2012.06595.x
Ann. N.Y. Acad. Sci. 1267 (2012) 95–102 © 2012 New York Academy of Sciences.

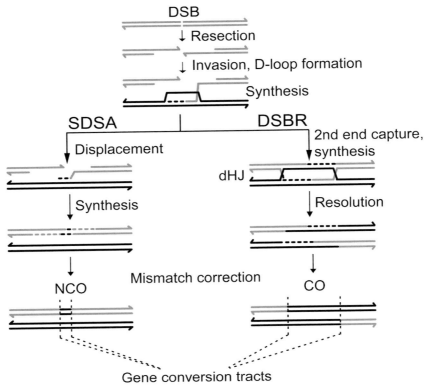

Figure 1. Meiotic recombination pathways. DSBs are induced preferentially at hotspots located throughout the genome. Resection of the DSB from 5′ to 3′ generates 3′ single-stranded tails, which invade the intact homolog (black) creating a D-loop intermediate. DNA repair synthesis (dashed lines) is primed by the invading 3′ end and templated by the homolog. In synthesis-dependent strand annealing (SDSA), the newly synthesized 3′ tail is displaced from the homolog, whereupon it anneals to the second, homologous 3′ end of the DSB. Subsequent repair synthesis and ligation reseals the DSB and generates a noncrossover (NCO). In double-strand break repair (DSBR), the D-loop captures the second end of the DSB and a double Holliday junction (dHJ) is formed. Resolution of the dHJ can generate a crossover (CO). Both NCOs and COs can result in gene conversions, as indicated, where sequences of the homolog that receives a DSB (gray) are converted to the genotype of the intact homolog (black).

which is used in mitosis. Based upon studies primarily performed in budding yeast, there are two major homologous recombination pathways that bifurcate from the D-loop intermediate. In one pathway, known as synthesis-dependent strand annealing (SDSA), the extended 3′ end of the invading strand is displaced after repair synthesis from the homolog to anneal to the second 3′ end on the other side of the DSB.[7] SDSA is thought to generate non-crossovers (NCOs) exclusively, which is a patch-like repair with no exchange of flanking markers. In the other pathway, known as double-strand break repair (DSBR), the second 3′ end is captured by the D-loop and a double Holliday junction (dHJ) forms.[8] dHJs can be resolved by structure-specific endonucleases; most such resolution events are thought to generate a crossover (CO), which is a reciprocal exchange

of flanking markers. Each homolog pair requires at least one CO to be physically linked as a bivalent, and these COs, in conjunction with sister chromatid cohesion established during DNA replication, tether the four chromatids (two from each homolog) as a tetrad.[2] Thus, COs provide the connections necessary for accurate reductional segregation in meiosis I.

How meiosis contributes to genetic diversity

Meiosis generates genetic diversity through three principal mechanisms. First, pairs of homologous chromosomes are independently assorted from each other into haploid gametes, for $2n$ possible combinations, where n is the number of homolog pairs. Second, CO recombination reciprocally exchanges chromosome segments between homologs, altering the association of maternal and paternal alleles at the point of exchange. Finally, gene conversion during homologous recombination results in non-Mendelian transmission of genetic information at the site of DSB repair (Fig. 1). If the region around a DSB is polymorphic between homologs, heteroduplex DNA can form during homologous recombination and mismatch correction will convert those polymorphisms. Thus, within the meiotic tetrad, information is transferred unidirectionally from one parental chromatid to another, resulting in 3:1 transmission of alleles within the gene conversion tract. While the contribution of independent assortment and CO recombination to genomic diversity has long been appreciated, the contribution of gene conversion, particularly as a result of NCOs, has been difficult to assess.

Toward a comprehensive assessment of meiotic gene conversion

COs are a minor outcome of homologous recombination in mammalian meiosis. Based upon cytological markers, COs likely represent only ~10% of DSB repair products in mouse spermatocytes.[9] The remaining DSBs had been inferred to be repaired as interhomolog NCOs; however, when analyzed, the proportion of NCOs at recombination hotspots was much lower in mammals than expected.[10–12] This finding raised the possibility that many mammalian meiotic DSBs could be repaired by other mechanisms, such as repair between sister chromatids or even by nonhomologous processes. However, repair of this nature would serve no obvious purpose in promoting homolog recognition, pairing, and tethering. An important technical consideration is that COs can always be identified because they result in the exchange of flanking markers, but NCOs can only be detected if they generate a gene conversion incorporating a polymorphism. This raised the question as to whether the low frequency of NCOs identified in previous studies could be due to the low density of polymorphisms in the analyzed recombination hotspots.

To counter this problem, we sought to characterize meiotic recombination at a highly polymorphic mouse hotspot, hypothesizing that even if NCO gene conversion tracts are very short, we would capture a substantial fraction of events.[13] We summarize our findings below and discuss some of the implications.

Frequency and distribution of COs and NCOs at a mouse hotspot: variations between strain backgrounds

We determined that a previously identified mouse hotspot termed *A3* (Ref. 14) had a high density of polymorphisms between inbred *Mus musculus* strains, ranging from 1.6% to 1.8%, with polymorphisms located approximately every 30 bp in the center of the hotspot. We compared the distribution of COs and NCOs in sperm from F1 hybrid mice from different strain combinations. Recombinant DNA molecules were amplified using allele-specific PCR, according to previously developed methods.[15,16] The *A3* hotspot was active for CO recombination in all F1 hybrids analyzed with a CO frequency of $\sim 10^{-3}$ per sperm genome, ~200-fold greater than the genome average. In A/J × DBA/2J F1 hybrids, the observed distribution of CO exchange points differed depending on which orientation of CO product was amplified from sperm DNA: When amplifying molecules in the DBA/2J to A/J orientation, most CO exchange points (i.e., where the amplified DNA sequence switches from the DBA/2J to A/J genotype) clustered to the left (Fig. 2A, top), but when amplifying molecules in the A/J to DBA/2J orientation, most CO exchange points clustered to the right (Fig. 2A, bottom). This asymmetric distribution pattern of CO exchange points has been proposed by Jeffreys *et al.* to be caused by preferential formation of DSBs on one chromosome compared to the other,[17] as the broken chromatid copies genetic information from the unbroken homolog (Fig. 1). Based on this model, we infer that DSBs at *A3* are to be strongly biased in favor of the DBA/2J chromosome compared with the A/J chromosome. As a result, CO gene conversion strongly favors transmission of information from the unbroken A/J chromosome. By determining how offset the distribution of exchange points is between orientations, we estimated that the mean CO gene conversion tract length at *A3* is ~500 bp (Fig. 2A). This estimate is similar to those at the handful of human and mouse hotspots that show similar asymmetric CO patterns.[17–20] Consistent with preferential DSB formation on the DBA/2J chromosome leading to biased NCO gene conversion (Fig. 1), approximately nine-fold more

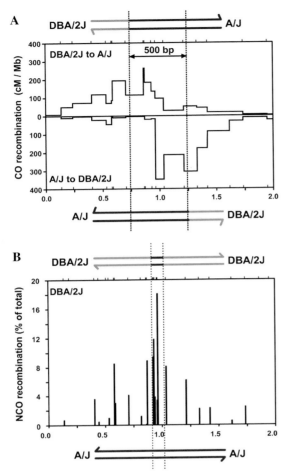

Figure 2. Recombination at the *A3* hotspot. (A) CO activity in centimorgans (cM) per Mb in A/J × DBA/2J F1 hybrids. When amplifying COs in the DBA/2J to A/J orientation, exchange points cluster to the left (top), while in the A/J to DBA/2J orientation, exchange points cluster to the right (bottom). The offset in exchange points is due to preferential DSB formation on the DBA/2J chromosome. Dashed lines indicate the distribution center of each orientation, and the offset between them is used to estimate the mean CO gene conversion tract length of 500 bp. (B) NCOs (as % of total NCOs detected) in all F1 hybrids on the DBA/2J chromosome. Ticks at the top of graphs represent tested polymorphisms. The *x*-axis scale for (A) and (B) is in kilobases.

NCOs were observed on the DBA/2J chromosome compared with the A/J chromosome.

In contrast to A/J × DBA/2J F1 hybrids, C57BL/6J × DBA/2J F1 hybrids showed a symmetric distribution pattern of CO exchange points at the *A3* hotspot:[13] COs clustered at the same point when amplified in either the C57BL/6J to DBA/2J or DBA/2J to C57BL/6J orientation. Moreover, NCOs

were similar in frequency for both the C57BL/6J and DBA/2J chromosomes. Thus, DSB formation appears to be equally frequent on each chromosome in this F1 hybrid strain background.

High-resolution analysis of NCO gene conversions

NCO gene conversions peak at the same central polymorphisms that are at the center of CO exchange points (compare Fig. 2A and B). In comparison to COs, however, NCO gene-conversion tracts were universally short. We determined that the median of the maximal NCO gene-conversion tracts in the center of the hotspot was ~100 bp, or approximately five-fold shorter than the mean of CO gene conversion tracts. Most NCO gene conversions incorporated only a single polymorphism, validating our hypothesis that a high polymorphism density is required to capture a significant fraction of NCOs. The greater sensitivity of detection of NCOs compared with earlier studies allowed us to determine that there are substantially more NCOs than COs at *A3*. For example, in the A/J × DBA/2J F1 hybrid, the NCO frequency is ~2.3% on a per meiosis basis, which is 10-fold higher than the CO frequency (0.22%). Thus, for the first time, the ratio of NCOs to COs at a hotspot approximates the cytological estimate of the ratio of global DSBs to COs observed in mouse spermatocytes.[9] The high NCO to CO ratio supports the hypothesis that most meiotic DSBs provoke interactions between homologs, which are critical for meiotic progression, in particular homolog pairing.

NCOs peaked in number at the center of the *A3* hotspot; however, a substantial proportion of NCOs occurred in the flanking regions of the hotspot. Considering NCOs on the DBA/2J chromosome for all of the strain combinations analyzed, only about half of NCOs localized within the central 43 bp of the *A3* hotspot (Fig. 2B). The remaining NCO gene conversions were located in the flanking ~750 bp on either side. As NCO gene conversions are short, we infer that the distribution of NCOs approximates the distribution of DSBs. Thus, DSBs do not form exclusively at the center of *A3*, but rather span a broad region encompassing ~1.5 kb. Direct analysis of meiotic DSBs in yeast[21–23] demonstrates that DSBs are distributed throughout the width of hotspots, and also frequently spread into the flanking regions, as inferred for the *A3* hotspot. If NCO gene conversion patterns at other mammalian hotspots are similar to what we observe at *A3* and what has been shown in yeast, then meiotic DSBs are clearly not restricted to the center of hotspots, as is frequently modeled in mammals (e.g., Refs. 24 and 25); thus, while knowing the exact center of a hotspot is useful,[26] it provides only part of the information regarding DSB distribution and the resultant gene conversion patterns.

Transmission distortion and allelic shuffling due to gene conversion: NCO supremacy?

Meiotic recombination can have a major impact on genomic architecture. COs break up allelic blocks on chromosome-wide scales, but gene conversion at hotspots from COs and NCOs also contribute to allelic shuffling, albeit over much smaller distances. Analyses of genetic maps in humans indicate that the major impact of meiotic recombination on genetic diversity occurs at the scale of hotspots, implicating gene conversion as a main driver of diversity.[27] Although NCO gene conversion tracts are smaller than those of COs, the contribution of NCOs can be significant owing to their high frequency. While it is formally possible that *A3* has an unusually high NCO to CO ratio, the concordance with the cytological ratio of DSBs to COs suggests that *A3* may be typical of mammalian hotspots.

Importantly, gene conversion can also lead to transmission distortion, that is, deviation from the expected 50:50 gametic ratio of parental alleles. Transmission distortion in this case is a consequence of preferential DSB formation on one homolog leading to overtransmission of alleles from the other, uncut homolog. The impact of gene conversion on transmission distortion depends on several variables: the absolute DSB frequency at the hotspot, the magnitude of DSB bias for one homolog over the other, and the length of the subsequent gene conversion tract. While transmission distortion due to CO gene conversion has been analyzed at several recombination hotspots in mouse and human,[17–20] the ability to capture a significant fraction of NCOs at *A3* provided an opportunity to assess the contribution of both CO and NCO gene conversion to transmission distortion. The percent transmission of the A/J allele peaks at ~70% for the polymorphism at the center of *A3* when considering all recombination events (black circle, Fig. 3A). When considering all gametes, the transmission distortion from CO gene conversion translates into a gametic ratio of 50.044:49.956 for the A/J allele. However, because of the much higher frequency of NCOs in the center of the hotspot, adding the transmission distortion from NCO gene conversion increases this gametic ratio to 50.23:49.77. If the polymorphism at the center of the hotspot also causes the DSB bias between A/J and DBA/2J, population simulations predict that transmission distortion will lead to fixation of the A/J polymorphism and hotspot quiescence (i.e., a dampening of hotspot activity) in less than 1,200 generations (black circles, Fig. 3B).

Similar simulations predict that the high frequency of recombination and the strong DSB preference for DBA/2J would also lead to fixation of other

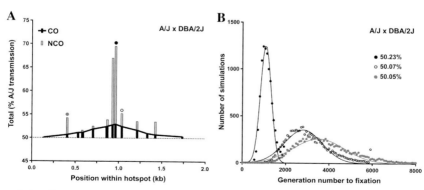

Figure 3. Transmission distortion and allelic fixation at the *A3* hotspot. (A) Percent A/J transmission for A/J × DBA/2J F1 hybrids from COs (black bars and line) and NCOs (gray bars). The dotted line at 50% is Mendelian transmission. Circles above the bars demarcate polymorphisms analyzed in B. (B) Monte Carlo simulations (Wright-Fisher model) to determine the number of generations to fixation for the three indicated polymorphisms, with the simplified assumption that the polymorphism is causative for the DSB bias. Calculated gametic ratios are indicated.

nearby polymorphisms in the *A3* hotspot. Polymorphisms just adjacent to the center of the hotspot may experience a more equitable contribution of NCOs and COs to transmission distortion (open circle, Fig. 3), whereas polymorphisms more distant from the center may undergo transmission distortion largely driven by NCOs (gray circle, Fig. 3). Thus, while short NCO gene conversion tracts have a more punctate effect on transmission distortion than COs, the substantially higher ratio of NCOs to COs at *A3* results in a 1.9-fold greater contribution of NCOs to transmission distortion (Fig. 3A). We predict that this greater contribution of NCOs would be observed genome wide, as the approximately five-fold longer CO gene conversion tracts are counterbalanced by ~10-fold more NCOs. Thus, NCOs are predicted to have a significant impact on allele fixation, hotspot quiescence, and, ultimately, genomic diversity.

Short NCO gene conversion tracts in mammals: implications for hotspot longevity

The breadth of NCO gene conversions across the *A3* hotspot also has consequences for the longevity of hotspots. Recent studies have determined that the location of DSB hotspots is largely controlled in mouse and humans by PRDM9, a meiosis-specific histone H3 methyltransferase.[28] PRDM9 has an array of Zn-finger DNA-binding motifs that target sites for SPO11-dependent DSBs.[29,30] The Zn-finger domain of PRDM9 is rapidly evolving,[31,32] suggesting that PRDM9 has undergone positive selection to alter its DNA-binding residues. The rapidly evolving DNA binding domain of PRDM9 may function to counteract the consequence of gene conversion on hotspot loss,[33] referred to as the *hotspot paradox*. This paradox posits that during meiosis any homolog that preferentially undergoes a DSB (the "hotter" allele) is undertransmitted compared to the "colder" allele of the other homolog. In this manner, hotspots would rapidly extinguish themselves, prompting the question: how can hotspots exist? Rapid evolution of PRDM9 DNA-binding sites acts to alter hotspot locations genome wide, providing a potential mechanism to counter hotspot loss.

Features of NCOs themselves nonetheless act to mitigate the speed of hotspot extinction. Biochem-

ical data indicate that PRDM9 preferentially binds to a DNA sequence at the center of hotspots,[30] and genetic data suggest that a single nucleotide polymorphism can alter PRDM9 binding.[34] The short conversion tracts of NCOs and the wide distribution of NCOs throughout the *A3* hotspot make it likely that a significant fraction of NCO gene conversion tracts do not span the PRDM9-binding site. In contrast, we infer from the pattern of CO exchange points that the majority of CO gene conversions would involve the PRDM9-binding site. The combination of the high NCO to CO ratio, short gene conversion tracts of NCOs, and the wide distribution of NCOs implies that many DSBs can promote homolog pairing without contributing to hotspot quiescence. In this manner, even extremely active hotspots can have longevity over evolutionary time scales.

Mammalian homologous recombination mechanisms are analogous in many ways to those in budding yeast. For example, many of the enzymes and structural proteins that mediate meiotic recombination are conserved. However, clear mechanistic differences exist between mammals and yeast, which can have profound consequences. Yeast does not have a PRDM9-dependent mechanism to target sites for SPO11 cleavage; instead, hotspots are, for the most part, associated with promoters of genes (see Ref. 23 and references therein). Further, yeast DSB hotspots are smaller in width, with a median of less than 200 bp, while mammalian hotspots, like *A3*, are wider, spanning ~1.5–2 kb. In yeast, COs outnumber NCOs, and gene conversion tracts are substantially longer, averaging ~2 kb for COs and 1.8 kb for NCOs.[35] If gene conversion mechanisms in mammals were identical to those in yeast, repair of most DSBs would convert the PRDM9-binding site, dramatically increasing the speed of hotspot quiescence.

The localization of hotspots in budding yeast to promoters means that hotspot sequences are evolutionarily constrained for reasons unrelated to meiotic recombination.[36] Hotspots are short lived in mammals—for example, hotspots in chimps differ in location from those in humans[31]—so clearly their location is not highly constrained. However, prolonged longevity of sequences that are favorable for DSB formation in mammals may also be found advantageous for as yet undetermined reasons.

Conclusion

Our comprehensive analysis of the *A3* hotspot has provoked new insights for how gene conversion can influence genomic diversity and evolution. Approximately 90% of meiotic recombination products at *A3* are NCOs, approaching for the first time the frequency predicted based upon cytological determination of the numbers of DSBs versus COs and providing support for the model that frequent DNA repair interactions promote homolog pairing during meiosis. Despite the more punctate effect of NCO gene conversion upon transmission distortion, the high frequency of NCOs results in approximately two-fold greater contribution to transmission distortion than COs. While half of NCOs occur in the center of the hotspot, the remaining NCOs are dispersed across the width of the hotspot. Due to the short gene conversion tracts associated with NCOs, NCOs in the flanking regions can promote homolog pairing and result in allelic shuffling without causing hotspot quiescence. Thus, many of the features of mammalian recombination—a high ratio of NCOs to COs, short gene conversion tracts, and wide hotspots—promote increased longevity of mammalian hotspots over evolutionary time scales.

Acknowledgments

We thank Aaron Gabow and Alex Lash of the MSKCC Bioinformatics Core for calculations of fixation and Liisa Kauppi and Robert J. Klein for helpful discussions. This work was supported by NIH grants HD040916 and HD53855 (M.J. and S.K.).

Conflicts of Interest

The authors declare no conflicts of interest.

References

1. Cole, F., S. Keeney & M. Jasin. 2010. Evolutionary conservation of meiotic DSB proteins: more than just Spo11. *Genes. Dev.* **24:** 1201–1207.
2. Zickler, D. & N. Kleckner. 1999. Meiotic chromosomes: integrating structure and function. *Annu. Rev. Genet.* **33:** 603–754.
3. Kauppi, L., M. Jasin & S. Keeney. 2012. The tricky path to recombining X and Y chromosomes in meiosis. *Ann. N.Y. Acad. Sci.* **1267:** 18–23. This volume.
4. Hassold, T., H. Hall & P. Hunt. 2007. The origin of human aneuploidy: where we have been, where we are going. *Hum. Mol. Genet.* **16:** R203–R208.
5. Hunter, N. 2007. *Meiotic Recombination, in Topics in Current Genetics, Molecular Genetics of Recombination*, Vol. 17/2007.
    A. Aguilera & R. Rothstein, Eds.: 381–442. Springer-Verlag. Heidelberg.
6. Keeney, S. 2007. *Spo11 and the Formation of DNA Double-Strand Breaks in Meiosis, in Recombination and Meiosis*. R. Egel & D.-H. Lankenau, Eds.: 81–123. Springer-Verlag. Berlin, Heidelberg.
7. Paques, F. & J.E. Haber. 1999. Multiple pathways of recombination induced by double-strand breaks in Saccharomyces cerevisiae. *Microbiol. Mol. Biol. Rev.* **63:** 349–404.
8. Szostak, J.W., T.L. Orr-Weaver, R.J. Rothstein & F.W. Stahl. 1983. The double-strand-break repair model for recombination. *Cell* **33:** 25–35.
9. Cole, F. *et al.* 2012. Homeostatic control of recombination is implemented progressively in mouse meiosis. *Nat. Cell Biol.* **14:** 424–430.
10. Guillon, H., F. Baudat, C. Grey, R.M. Liskay & B. de Massy. 2005. Crossover and noncrossover pathways in mouse meiosis. *Mol. Cell* **20:** 563–573.
11. Jeffreys, A.J. & C.A. May. 2004. Intense and highly localized gene conversion activity in human meiotic crossover hot spots. *Nat. Genet.* **36:** 151–156.
12. Jeffreys, A.J., R. Neumann, M. Panayi, S. Myers & P. Donnelly. 2005. Human recombination hot spots hidden in regions of strong marker association. *Nat. Genet.* **37:** 601–606.
13. Cole, F., S. Keeney & M. Jasin. 2010. Comprehensive, fine-scale dissection of homologous recombination outcomes at a hot spot in mouse meiosis. *Mol. Cell* **39:** 700–710.
14. Kelmenson, P.M. *et al.* 2005. A torrid zone on mouse chromosome 1 containing a cluster of recombinational hotspots. *Genetics* **169:** 833–841.
15. Kauppi, L., C.A. May & A.J. Jeffreys. 2009. Analysis of meiotic recombination products from human sperm. *Methods Mol. Biol.* **557:** 323–355.
16. Cole, F. & M. Jasin. 2011. Isolation of meiotic recombinants from mouse sperm. *Methods Mol. Biol.* **745:** 251–282.
17. Jeffreys, A.J. & R. Neumann. 2002. Reciprocal crossover asymmetry and meiotic drive in a human recombination hot spot. *Nat. Genet.* **31:** 267–271.
18. Baudat, F. & B. de Massy. 2007. Cis- and trans-acting elements regulate the mouse Psmb9 meiotic recombination hotspot. *PLoS Genet.* **3:** e100.
19. Bois, P.R. 2007. A highly polymorphic meiotic recombination mouse hot spot exhibits incomplete repair. *Mol. Cell Biol.* **27:** 7053–7062.
20. Webb, A.J., I.L. Berg & A. Jeffreys. 2008. Sperm cross-over activity in regions of the human genome showing extreme breakdown of marker association. *Proc. Natl. Acad. Sci. USA* **105:** 10471–10476.
21. de Massy, B., V. Rocco & A. Nicolas. 1995. The nucleotide mapping of DNA double-strand breaks at the CYS3 initiation site of meiotic recombination in Saccharomyces cerevisiae. *EMBO J* **14:** 4589–4598.
22. Liu, J., T.C. Wu & M. Lichten. 1995. The location and structure of double-strand DNA breaks induced during yeast meiosis: evidence for a covalently linked DNA-protein intermediate. *Embo J.* **14:** 4599–4608.
23. Pan, J. *et al.* 2011. A hierarchical combination of factors shapes the genome-wide topography of yeast meiotic recombination initiation. *Cell* **144:** 719–731.

24. Calabrese, P. 2007. A population genetics model with recombination hotspots that are heterogeneous across the population. *Proc. Natl. Acad. Sci. USA* **104:** 4748–4752.

25. Pineda-Krch, M. & R.J. Redfield. 2005. Persistence and loss of meiotic recombination hotspots. *Genetics* **169:** 2319–2333.

26. Smagulova, F. *et al.* 2011. Genome-wide analysis reveals novel molecular features of mouse recombination hotspots. *Nature* **472:** 375–378.

27. Spencer, C.C. *et al.* 2006. The influence of recombination on human genetic diversity. *PLoS Genet.* **2:** e148.

28. Neale, M.J. 2010. PRDM9 points the zinc finger at meiotic recombination hotspots. *Genome Biol.* **11:** 104.

29. Segurel, L., E.M. Leffler & M. Przeworski. 2011. The case of the fickle fingers: how the PRDM9 zinc finger protein specifies meiotic recombination hotspots in humans. *PLoS Biol.* **9:** e1001211.

30. Grey, C. *et al.* 2011. Mouse PRDM9 DNA-binding specificity determines sites of histone H3 lysine 4 trimethylation for initiation of meiotic recombination. *PLoS Biol.* **9:** e1001176.

31. Myers, S. *et al.* 2010. Drive against hotspot motifs in primates implicates the PRDM9 gene in meiotic recombination. *Science* **327:** 876–879.

32. Oliver, P.L. *et al.* 2009. Accelerated evolution of the Prdm9 speciation gene across diverse metazoan taxa. *PLoS Genet.* **5:** e1000753.

33. Coop, G. & S.R. Myers. 2007. Live hot, die young: transmission distortion in recombination hotspots. *PLoS Genet.* **3:** e35.

34. Jeffreys, A.J. & R. Neumann. 2005. Factors influencing recombination frequency and distribution in a human meiotic crossover hotspot. *Hum. Mol. Genet.* **14:** 2277–2287.

35. Mancera, E., R. Bourgon, A. Brozzi, W. Huber & L.M. Steinmetz. 2008. High-resolution mapping of meiotic crossovers and non-crossovers in yeast. *Nature* **454:** 479–485.

36. Nicolas, A., D. Treco, N.P. Schultes & J.W. Szostak. 1989. An initiation site for meiotic gene conversion in the yeast Saccharomyces cerevisiae. *Nature* **338:** 35–39.

Ann. N.Y. Acad. Sci. ISSN 0077-8923

ANNALS OF THE NEW YORK ACADEMY OF SCIENCES
Issue: *Effects of Genome Structure and Sequence on Variation and Evolution*

Gross chromosomal rearrangement mediated by DNA replication in stressed cells: evidence from *Escherichia coli*

J.M. Moore,[1] Hallie Wimberly,[2,3] P.C. Thornton,[2] Susan M. Rosenberg,[1,2,4,5] and P.J. Hastings[2]

[1]Department of Biochemistry and Molecular Biology, [2]Department of Molecular and Human Genetics, Baylor College of Medicine, Houston, Texas. [3]Department of Pathology, Yale University, New Haven, Connecticut. [4]Department of Molecular Virology and Microbiology, [5]Dan L. Duncan Cancer Center, Baylor College of Medicine, Houston, Texas

Address for correspondence: P.J. Hastings, Department of Molecular and Human Genetics, Baylor College of Medicine, 1 Baylor Plaza, Houston, TX 77030. hastings@bcm.edu

Gross chromosomal rearrangements (GCRs), or changes in chromosome structure, play central roles in evolution and are central to cancer formation and progression. GCRs underlie copy number variation (CNV), and therefore genomic disorders that stem from CNV. We study amplification in *Escherichia coli* as a model system to understand mechanisms and circumstances of GCR formation. Here, we summarize observations that led us to postulate that GCR occurs by a replicative mechanism as part of activated stress responses. We report that we do not find RecA to be downregulated by stress on a population basis and that constitutive expression of RecA does not inhibit amplification, as would be expected if downregulation of RecA made cells permissive for nonhomologous recombination. Strains deleted for the genes for three proteins that inhibit RecA activity, *psiB*, *dinI*, and *recX*, all show unaltered amplification, suggesting that if they do downregulate RecA indirectly, this activity does not promote amplification.

Keywords: nonhomologous recombination; amplification; stress; copy number variation; stress-induced mutation; genome rearrangement

Gross chromosomal rearrangements (GCRs) underlie many aspects of evolution, from short-term modulation of levels of gene expression to the formation of new functions by shuffling exons or domains from other genes, reassorting genes and regulatory elements, and providing genetic redundancy that allows progressive change in sequence. The discovery that GCR can occur as a component of stress responses[1,2] opens a wealth of possibilities for processes of adaptive evolution occurring specifically when organisms are not well-adapted to their environments, as, for example, during times of changing environment. Stress-inducible genomic instability has been described in several bacterial systems, in yeast, and in human cell cultures, each displaying slightly different characteristics in different assay systems but sharing requirements for activation of stress responses that in turn activate genome-instability pathways (for review, see Refs. 3–5). GCRs also underlie genomic disorders (disease syndromes that stem from variation in gene copy number), and the origin of some, and progression of, many cancers.[6] Thus, the mechanisms that underlie GCR are of broad interest in both basic science and human health.

We have studied starving *Escherichia coli* using the Lac assay,[7] a model organism assay system in which gene amplification, a GCR, occurs in response to the stress of starvation in the presence of a potential carbon source that the organism is unable to use. Amplification in this assay depends on the activation of two stress responses.[2,8] This report outlines the molecular processes that we have found to be involved, and the molecular mechanisms that these data suggest. We include data that exclude some possible explanations as to why cells undergo nonhomologous, instead of homologous, recombination (HR) during stress. The replication-based molecular mechanisms that we have proposed find support in observation of GCR in other organisms, notably in human copy number variation (CNV) encountered in the genetics clinic.

doi: 10.1111/j.1749-6632.2012.06587.x

Stress-induced mutation in starving *E. coli*

The model system

In the *E. coli* Lac assay,[7] a +1 frameshift mutation in a *lacI* fusion gene on an F′ plasmid mutates to Lac$^+$ during prolonged starvation on lactose minimal medium. The Lac$^+$ colonies carry either a compensating frameshift mutation[9,10] or a tandem array of 20 or more copies of the weakly functional *lac* allele, which confers sufficient β-galactosidase activity for growth.[1] The processes of frameshift (point) mutation and amplification differ in their genetic requirements, and thus represent alternative strategies that allow escape from starvation.

Requirements for stress-induced point mutation

Formation of point mutations differs from mutation formation in growing cells in its requirement for the proteins of homologous recombinational double-strand end (DSE) repair (RecA, RecBC, and RuvABC),[11–13] the SOS DNA-damage response,[14] error-prone DNA polymerase (Pol) IV/DinB,[15,16] the RpoS (σ^S) general-stress response,[2,17] and periplasmic-stress response controlled by σ^E.[8] Point mutation formation also requires the F-encoded TraI endonuclease– or I-SceI endonuclease–induced DSEs near *lac*,[18] though recent work shows that they also occur at spontaneous DSEs in chromosomes of F$^-$ *E. coli*.[19] TraI appears to provide ssDNA nicks that become DSEs by replication fork collapse. Hence, point mutations are thought to arise via DinB/Pol IV errors during DNA replication reinitiated by HR at collapsed replication forks.[18,19]

Stress-induced amplification

Amplification can rescue leaky mutants

Lac$^+$ colonies also arise in the Lac assay by gene amplification. Amplification is also an adaptive change that occurs after starvation has begun.[1] Amplification of the leaky *lac* allele gives Lac$^+$ colonies because 20 or more copies of the frameshift mutant gene provide sufficient β-galactosidase activity to allow growth on lactose medium. *lac*-amplified colonies are distinguished from point mutants by their instability, seen as blue and white sectoring on rich medium with X-gal, because amplification breaks down by HR between the repeats.[20,21] The repeat units (amplicons) are direct repeats of ~7 to ≥40 kb joined by microhomology junctions of 2–15 base pairs.[22,23]

Protein requirements for stress-induced amplification

Like point mutation, amplification requires activation of the RpoS and RpoE stress responses,[2,8] the proteins of DSE-repair (which might be a requirement for HR for expansion of an amplified array by unequal crossing-over from a duplication).[23] However, amplification also requires DNA Pol I, which is not required for stress-induced point mutation.[23,24] Amplification is enhanced by mutation in *xonA* (which encodes the major single-strand 3′-exonuclease ExoI),[23] and by providing DSEs with an endogenously expressed I-SceI endonuclease,[18] implying that 3′ ends at DSEs are intermediates in amplification. Amplification does not require DinB or the SOS response.[15]

Mechanism of stress-induced amplification

We proposed that amplification is initiated by template switching during repair of collapsed replication forks in cells unable to use HR,[23,25] a model that is currently important in human CNV work (e.g., Refs. 26–28). This model is described later.

The elevated rate of amplification seen in ExoI-defective mutants implies that 3′ single-strand DNA ends are intermediates in the amplification process, and that these ends are frequently removed by ExoI so that amplification is inhibited. Because Pol I functions in excision repair processes and in lagging-strand processing during replication, we tested for a requirement for excision repair in amplification. We found that nucleotide excision repair and base excision repair are not required,[23] nor is mismatch repair (Fig. 1). Thus, the requirement for Pol I implies that the events occurred during DNA replication at the replication fork. The presence of microhomology at the novel junctions of amplicons indicates that the initial event involved nonhomologous recombination (NHR), because the length of microhomology is too short to allow for RecA-mediated HR.[29]

An event involving single-strand 3′ DNA-ends at replication forks suggests that the novel junctions are formed by polymerase slippage or template switching during DNA replication. Template switching, as had been described previously, occurs within the boundaries of a replication fork. However, amplicons in the Lac assay system average about 20 kb in length,[23] presumably much too long to have occurred by polymerase switching within a single replication fork. We initially suggested that

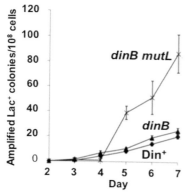

Figure 1. Mismatch repair is not required for amplification. The figure shows a standard stress-induced amplification experiment using the Lac assay. Methods were as published.[23] This plot (and others below (Fig. 3)) shows the accumulation of amplified Lac+ colonies on minimal lactose medium over time (days). The graph shows the mean with SEM of three or four independent cultures. All experiments have been performed at least three times with comparable results. To test whether deletion of *mutL* affected amplification, we used a *dinB mutL* double mutant. This reduces the amount of point mutation (see Ref. 15), which is very high in *mutL* cells but, as can be seen in the figure, does not affect amplification. Amplification is increased in *dinB* by the absence of MutL, demonstrating that MutL is not required and that, therefore, the requirement for Pol I in amplification is not a requirement for mismatch repair. All strains used were in an FC40 (Ref. 7) background and were constructed as previously published (e.g., Ref. 23): Din+, SMR4562; *dinB10*, SMR5830, and Δ*mutL dinB10*, SMR7956.

template switching occurred between replication forks: the long-distance template switch model.[23] This model was encouraged by the observation that some amplification events were complex, having a sequence from nearby regions inserted into the junction in either orientation.[23] Since then, the same observation has been made for human CNVs,[30] suggesting a common mechanism.[23,30]

A subset of cells experiences amplification

An extensive analysis of the structure of amplicons in the Lac assay by comparative genomic hybridization characterizes this complexity in more detail.[31] It also provides evidence of genome-wide instability, witnessed by a significant number of GCR events other than *lac* amplification occurring in the same cells that had experienced amplification, when compared with cells that have not been stressed, as well as with cells in the same stressed population but in which *lac* was not found to be amplified.

Genome-wide instability is predicted if amplification occurs as a consequence of a stress response because the response is a cell-wide phenomenon. We found evidence of genome-wide instability, but only in those cells that also carried amplification, confirming that a cell-wide physiological change underlies stress-induced amplification. At the same time, because only cells in which *lac* is amplified show additional GCRs, we infer that amplification occurs in a subpopulation of cells that is differentiated to be permissive for NHR.[31]

About 15% of amplification events were complex, with a mixture of direct and inverted insertions.[31] We now interpret this complexity as a product of break-induced replication (BIR), a process by which collapsed (broken) replication forks are repaired and restarted.[25,32] BIR has been shown in yeast to involve repeated rounds of extension by replication of a DSE followed by separation from the template and then reinvasion of the DNA end into new template DNA and priming of replication. This happens several times before a fully processive replication fork is established.[33] However, BIR is an HR-mediated process, whereas amplification in the Lac system occurs at microhomologous positions. We have therefore suggested that in stressed cells HR is not available, and instead, BIR occurs by annealing of the 3′ DNA-end with any nearby single-stranded DNA with which it shares microhomology.[25,32] This model is discussed later.

Microhomology-mediated BIR

We suggest that GCRs are formed by a modified BIR process (Fig. 2).[25,32] Because BIR is a precisely homologous process mediated by RecA (or its orthologue Rad51 in yeast),[34,35] we postulated that a failure to form homologous junctions in amplification could be due to an insufficiency of RecA activity in starving cells. Alternatively, it might result instead from cells having no sister chromosome at the time of repair, as is expected in 60% of stationary-phase cells.[36] To repair a collapsed fork without RecA or homology, one would need to use annealing of ssDNA. One strand is provided by the 3′-end from the processed DNA end at the site of fork collapse. We postulate that the ssDNA end pairs with any other ssDNA nearby. SsDNA is likely to occur, for example, at replication forks, excision repair sites, R-loops formed at sites of transcription, and at secondary structures in DNA, such

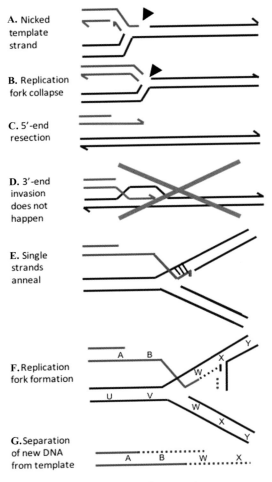

A. Nicked template strand

B. Replication fork collapse

C. 5'-end resection

D. 3'-end invasion does not happen

E. Single strands anneal

F. Replication fork formation

G. Separation of new DNA from template

C, E, F and G are repeated until a processive replication fork is formed

Figure 2. The MMBIR model. Lines indicate strands of DNA, arrow heads indicate 3' ends, and dotted lines show new DNA synthesis. Panel A shows a replication fork approaching a nick in a template strand. In B, one arm of the fork broke off when the fork reached the nick, resulting in a collapsed replication fork. (C) 5'-end resection produces a 3' single-stranded tail. (D) The expected sequel, that the 3'-end invades homologous duplex DNA, is postulated not to happen. (E) Instead, the 3'-end anneals with any nearby ssDNA, here shown as the leading-strand template of another replication fork. (F) A bidirectional replication fork is formed, and both strands are extended at the double-strand end. (G) The extended end separates from the template. Failing to find a double-strand end from the other side of the break (because there is none, as this is a collapsed fork), it will again find ssDNA and anneal, repeating steps E, F, and G a few times before a fully processive replication fork is formed. Thus, nonhomologous recombination occurs with microhomology junctions and complex insertions of a nearby sequence at the joint.

as hairpins and G-quartet DNA.[49] Because the homology requirements for strand annealing are so low, the 3' end could switch templates to almost any exposed ssDNA, and thus cause GCRs joined by microhomology, such as those that we see in amplification.[23]

Hence, the microhomology-mediated break-induced replication (MMBIR) model (Fig. 2) suggests that when a replication fork collapses in a cell under stress replication will be restarted by a modified BIR process in which HR functions are unavailable. Resection (exonuclease digestion) of the 5'-DNA end at the break produces a 3'-overhang (Fig. 2C). This 3'-end will anneal with any single-stranded DNA in physical proximity (Fig. 2E) and then prime synthesis (Fig. 2F). This annealing reaction has very low homology requirements, so that replication recommences (Fig. 2F) at a sequence that shows only microhomology, and almost any ssDNA sequence will be able to take part. This sequence can be in a region already replicated, producing a duplication, or downstream of where the fork collapsed, leading to deletion. Because the available ssDNA will often be a lagging-strand template, inversion will be frequent. As in BIR, initial synthesis is of low processivity, and the extended end will separate from the template (Fig. 2G). The process must then be repeated to produce a completed viable chromosome, so that complexity at the joints will occur in the form of an inserted sequence, often from nearby, in either orientation. After a few such repeated events, fully processive replication is established and continues to the end of the replicon. The final replication would usually be in direct orientation to obtain a viable product.

Other mechanisms for GCR besides template switching during replication must be considered. The most likely candidate is nonhomologous end-joining (NHEJ; reviewed in Refs. 37 and 38) and NHEJ has frequently been cited as a probable mechanism for human GCR. There is sometimes microhomology at junctions formed by NHEJ, and there is also a tendency for fragments of a sequence from elsewhere to be incorporated into the junction. In addition, insertions or deletions of one or two base pairs occur at some of the junctions, and we have observed this in *E. coli*.[23] However, *E. coli* lacks the necessary proteins for NHEJ,[39] and yet the properties of GCR in *E. coli* and in humans are very similar, suggesting a common mechanism that is

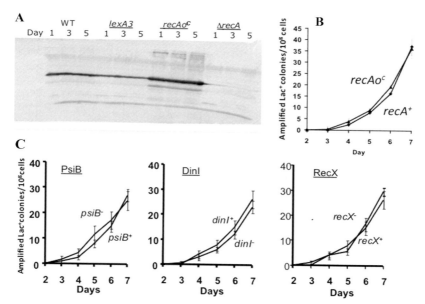

Figure 3. (A) Western blot of RecA in starving cells. There is no change in average RecA content of wild-type cells as they enter and stay in stationary phase (days 3 and 5), compared with the level in logarithmic phase (day 1). The cells were grown in M9 glycerol medium at 37 °C. Under these conditions, the cultures are in the logarithmic phase of growth on day 1 and in stationary phase on days 3 and 5. This is also seen in *lexA3* cells that are unable to induce the SOS DNA damage response. (B) High constitutive expression of RecA from the *recAo281* allele (shown in A) does not affect amplification, suggesting that amplification in *E. coli* does not result from downregulation of RecA protein by stress response. (C) Removal of proteins that inhibit RecA activity does not affect amplification. Deletion of *psiB, dinI,* or *recX* does not inhibit amplification, demonstrating that RecA activity in amplification is not controlled indirectly by these proteins. Strains used were wild-type, SMR4562; *lexA3,* SMR868; *recAo281,* PJH1139; Δ*recA*::Kan, PJH1146; Δ*psiB*::Cam, SMR5399; Δ*dinI2,* SMR4697, and Δ*recX,* PJH1101.

not NHEJ. Another argument against NHEJ is that the sequences inserted in both *E. coli* and humans generally reflect a sequence from the immediate genomic region rather than random fragments from elsewhere in the cell. Moreover, the requirement for Pol I[23,24] implies a replicative mechanism. Evidence for a replicative mechanism is also found in other recombination assays in *E. coli* (reviewed in Ref. 40) and yeast.[41]

What promotes NHR under stress?

Because BIR is so efficient and accurate there is a puzzle as to why NHR is allowed to occur, because the homology requirement for BIR would be expected to exclude nonhomologous and microhomologous events. If RecA and sequence homology were present, RecA would catalyze invasion of the homologous sequence and repair would be accurate. We suggest that RecA activity and/or homology is limiting in these starving cells. RecA/Rad51 could be downregulated by stress, as has been demonstrated in human cancer cell lines under hypoxic stress,[42,43]

during which (under stress) HR is also reduced and replaced by NHR. A switch from HR to NHR also occurs in *Drosophila* that are heterozygous for mutation in a Rad51 homologue, showing again that NHR happens when RecA/Rad51 activity is limiting.[44] However, we find no evidence of downregulation of RecA in starved *E. coli* (Fig. 3A), and a strain with high constitutive expression of RecA is unaltered for amplification (Fig. 3B). It is also possible that there is not enough adenosine-5′-triphosphate (ATP) for RecA activity in starving cells. Because, as discussed earlier, amplification occurs in a differentiated subpopulation of cells,[31] we must consider the possibility that differentiation to being permissive for NHR will apply only to a subpopulation. The Western blots presented in Figure 3 examine only the whole population, and so they could miss subpopulation-specific downregulation of RecA.

Another possibility for why repair might not use HR could be stress-response control of proteins that inhibit RecA activity, so that control of RecA activity would be indirect. However, as shown in

Figure 3C, three inhibitors of RecA activity, PsiB, DinI, and RecX (reviewed in Ref. 45), are not required for amplification. Other possibilities are that the absence of a sister molecule with which to interact (expected in 60% of cells[36]) and downregulation of DNA Pol III allowing access of Pol I,[46] which is more conducive to GCR. These experiments are not subject to concerns about subpopulations.

Conclusions

There is now extensive evidence supporting the idea that aberrant DNA replication underlies much GCR. Previous research in *E. coli*, yeast, and humans suggests that GCRs can be formed through the process of restarting collapsed replication forks by BIR. However, any BIR process that generates amplification in the Lac assay, and GCR in many other systems, does not use HR involving RecA/Rad51-mediated invasion of homologous duplex DNA to create novel junctions and duplications because the microhomology at the junctions is too short for HR. HR is necessary to expand a duplication into an amplified array, which probably underlies the requirement for HR proteins in amplification.[23] We suggest that, instead of RecA-mediated invasion, novel junction formation involves annealing of single stranded DNA at the broken end with any other single-stranded DNA nearby at sites of microhomology.[23,25] A key question is to understand the molecular mechanism by which stress changes cellular physiology such that nonhomologous events occur in situations where normally there would be HR; for example, collapsed replication forks usually are repaired using homologous sequences in the sister DNA molecule. We present data showing that, unlike human cells, *E. coli* in our conditions does not downregulate RecA/Rad51 under stress, at least when viewed on a whole population basis. Attempts to find indirect regulation of RecA activity under stress have not yielded an answer. Other ways to limit HR, for example, by the unavailability of a homologous sequence, are under consideration.

Chromosomal rearrangement is an evolutionary engine. It creates new regulatory circuits and new genes derived by reassortment of existing modules, expands the genome in a way that allows sequence diversity to evolve, and varies gene copy number, which can change expression levels. The discovery that stress responses promote amplification implies that the capacity of organisms to evolve can be increased specifically when they are maladapted to their environment, when stressed. This contrasts with early ideas of the neoDarwinian "modern synthesis" about constant and gradual genetic changes underlying evolution,[47] and instead implies feedback and responsiveness of the generation of genomic diversity to the environment. Further, that such accelerated genome evolution happens in only a subpopulation of cells suggests that the apparent danger of reshuffling a genome during stress could be mitigated by differentiating a small subpopulation in which to run the potentially "dangerous experiment". Both stress-induced GCR and stress-induced point mutation[3] may, on a microscopic scale, help contribute to the macroscopic phenomenon of evolution seeming to occur in bursts.[48]

Acknowledgments

This work was supported by NIH Grants R01 GM53158 to SMR and by R01 GM64022 to P.J.H.

Conflicts of interest

The authors declare no conflicts of interest.

References

1. Hastings, P.J. *et al.* 2000. Adaptive amplification: an inducible chromosomal instability mechanism. *Cell* **103:** 723–731.
2. Lombardo, M.-J., I. Aponyi & S.M. Rosenberg. 2004. General stress response regulator RpoS in adaptive mutation and amplification in *Escherichia coli. Genetics* **166:** 669–680.
3. Galhardo, R.S., P.J. Hastings & S.M. Rosenberg. 2007. Mutation as a stress response and the regulation of evolvability. *Crit. Rev. Biochem. Mol. Biol.* **42:** 399–435.
4. Hastings, P.J. 2007. Adaptive amplification. *Crit. Rev. Biochem. Mol. Biol.* **42:** 1–13.
5. Bindra, R.S. & P.M. Glazer. 2007. Repression of RAD51 gene expression by E2F4/p130 complexes in hypoxia. *Oncogene* **26:** 2048–2057.
6. Stratton, M.R., P.J. Campbell & P.A. Futreal. 2009. The cancer genome. *Nature* **458:** 719–724.
7. Cairns, J. & P.L. Foster. 1991. Adaptive reversion of a frameshift mutation in *Escherichia coli. Genetics* **128:** 695–701.
8. Gibson, J.L. *et al.* 2010. The σE stress response is required for stress-induced mutagenesis in *Escherichia coli. Mol. Micro.* **77:** 415–430.
9. Foster, P.L. & J.M. Trimarchi. 1994. Adaptive reversion of a frameshift mutation in *Escherichia coli* by simple base deletions in homopolymeric runs. *Science* **265:** 407–409.
10. Rosenberg, S.M. *et al.* 1994. Adaptive mutation by deletions in small mononucleotide repeats. *Science* **265:** 405–407.
11. Harris, R.S., S. Longerich & S.M. Rosenberg. 1994. Recombination in adaptive mutation. *Science* **264:** 258–260.

12. Harris, R.S., K.J. Ross & S.M. Rosenberg. 1996. Opposing roles of the Holliday junction processing systems of *Escherichia coli* in recombination-dependent adaptive mutation. *Genetics* **142:** 681–691.

13. Foster, P.L., J.M. Trimarchi & R.A. Maurer. 1996. Two enzymes, both of which process recombination intermediates, have opposite effects on adaptive mutation in *Escherichia coli*. *Genetics* **142:** 25–37.

14. McKenzie, G.J. *et al.* 2000. The SOS response regulates adaptive mutation. *Proc. Natl. Acad. Sci. U.S.A.* **97:** 6646–6651.

15. McKenzie, G.J. *et al.* 2001. SOS mutator DNA polymerase IV functions in adaptive mutation and not adaptive amplification. *Mol. Cell* **7:** 571–579.

16. Galhardo, R.S. *et al.* 2009. DinB upregulation is the sole role of the SOS response in stress-induced mutagenesis in *Escherichia coli*. *Genetics* **182:** 55–68.

17. Layton, J.C. & P.L. Foster. 2003. Error-prone DNA polymerase IV is controlled by the stress-response sigma factor, RpoS, in *Escherichia coli*. *Mol. Microbiol.* **50:** 549–561.

18. Ponder, R.G., N.C. Fonville & S.M. Rosenberg. 2005. A switch from high-fidelity to error-prone DNA double-strand break repair underlies stress-induced mutation. *Mol. Cell* **19:** 791–804.

19. Shee, C. *et al.* 2011. Impact of a stress-inducible switch to mutagenic repair of DNA breaks on mutation in Escherichia. coli. *Proc. Natl. Acad. Sci. U.S.A.* **108:** 13659–13664.

20. Tlsty, T.D., A.M. Albertini & J.H. Miller. 1984. Gene amplification in the *lac* region of *E. coli*. *Cell* **37:** 217–224.

21. Novick, A. & T. Horiuchi. 1961. Hyper-production of beta-galactosidase by *Escherichia coli* bacteria. *Cold Spring Harb. Symp. Quant. Biol.* **26:** 239–245.

22. Kugelberg, E. *et al.* 2006. Multiple pathways of selected gene amplification during adaptive mutation. *Proc. Natl. Acad. Sci. U.S.A.* **103:** 17319–17324.

23. Slack, A. *et al.* 2006. On the mechanism of gene amplification induced under stress in *Escherichia coli*. *PLoS Genetics* **2:** e48.

24. Hastings, P.J. *et al.* 2004. Adaptive amplification and point mutation are independent mechanisms: evidence for various stress-inducible mutation mechanisms. *PLoS Biol.* **2:** e399.

25. Hastings, P.J., G. Ira & J.R. Lupski. 2009. A Microhomology-mediated Break-Induced Replication Model for the origin of human copy number variation. *PLoS Genet.* **5:** e1000327.

26. Bedoyan, J.K. *et al.* 2011. A complex 6p25 rearrangement in a child with multiple epiphyseal dysplasia. *Am. J. Med. Genet. A* **155A:** 154–163.

27. Kidd, J.M. *et al.* 2010. A human genome structural variation sequencing resource reveals insights into mutational mechanisms. *Cell* **143:** 837–847.

28. Quemener, S. *et al.* 2010. Complete ascertainment of intragenic copy number mutations (CNMs) in the CFTR gene and its implications for CNM formation at other autosomal loci. *Hum. Mutat.* **31:** 421–428.

29. Lovett, S.T. *et al.* 2002. Crossing over between regions of limited homology in *Escherichia coli*. RecA-dependent and RecA-independent pathways. *Genetics* **160:** 851–859.

30. Lee, J.A., C.M. Carvalho & J.R. Lupski. 2007. A DNA replication mechanism for generating nonrecurrent rearrangements associated with genomic disorders. *Cell* **131:** 1235–1247.

31. Lin, D. *et al.* 2011. Global chromosomal structural instability in a subpopulation of starving *Escherichia coli* cells. *PLoS Genetics* **7:** e1002223.

32. Hastings, P.J. *et al.* 2009. Mechanisms of change in gene copy number. *Nat. Rev. Genet.* **10:** 551–564.

33. Smith, C.E., B. Llorente & L.S. Symington. 2007. Template switching during break-induced replication. *Nature* **447:** 102–105.

34. McEachern, M.J. & J.E. Haber. 2006. Break-induced replication and recombinational telomere elongation in yeast. *Annu. Rev. Biochem.* **75:** 111–135.

35. Morrow, D.M., C. Connelly & P. Hieter. 1997. "Break-copy" duplication: a model for chromosome fragment formation in *Saccharomyces cerevisiae*. *Genetics* **147:** 371–382.

36. Akerlund, T., K. Nordstrom & R. Bernander. 1995. Analysis of cell size and DNA content in exponentially growing and stationary-phase batch cultures of *Escherichia coli*. *J. Bacteriol.* **177:** 6791–6797.

37. Lieber, M.R. 2008. The mechanism of human nonhomologous DNA end joining. *J. Biol. Chem.* **283:** 1–5.

38. Lieber, M.R. 2010. The mechanism of double-strand DNA break repair by the nonhomologous DNA end-joining pathway. *Annu. Rev. Biochem.* **79:** 181–211.

39. Wilson, T.E., L.M. Topper & P.L. Palmbos. 2003. Nonhomologous end-joining: bacteria join the chromosome breakdance. *Trends Biochem. Sci.* **28:** 62–66.

40. Bzymek, M. & S.T. Lovett. 2001. Instability of repetitive DNA sequences: the role of replication in multiple mechanisms. *Proc. Natl. Acad. Sci. U.S.A.* **98:** 8319–8325.

41. Payen, C. *et al.* 2008. Segmental duplications arise from Pol32-dependent repair of broken forks through two alternative replication-based mechanisms. *PLoS Genet.* **4:** e1000175.

42. Bindra, R.S. *et al.* 2005. Alterations in DNA repair gene expression under hypoxia: elucidating the mechanisms of hypoxia-induced genetic instability. *Ann. N.Y. Acad. Sci.* **1059:** 184–195.

43. Bindra, R.S.S. *et al.* 2004. Downregulation of Rad51 and decreased homologous recombination in hypoxic cancer cells. *Mol. Cell Biol.* **24:** 8504–8518.

44. McVey, M. *et al.* 2004. Formation of deletions during double-strand break repair in Drosophila DmBlm mutants occurs after strand invasion. *Proc. Natl. Acad. Sci. U.S.A.* **101:** 15694–15699.

45. Cox, M.M. 2007. Regulation of bacterial RecA protein function. *Crit. Rev. Biochem. Mol. Biol.* **42:** 41–63.

46. Frisch, R.L. *et al.* 2010. There are separate DNA Pol II- and Pol IV-dependent components of double-strand-break-repair-associated stress-induced mutagenesis and RpoS controls both. *J. Bacteriol.* **192:** 4694–4700.

47. Mayr, E. 1982. *The Growth of Biological Thought, Diversity, Evolution, and Inheritance*. Belknap. Cambridge, MA.

48. Eldredge, N. & S.J. Gould. 1972. PUnctuated equilibrium: an alternative to phyletic gradualism. In *Models in Paleobiology*. T.J.M. Schopf, Ed.: 82–115. Freeman Cooper and Co. San Francisco.

49. Maizels, N. 2012. G4 motifs in human genes. *Ann. N.Y. Acad. Sci.* **1267:** 53–60. This volume.

Ann. N.Y. Acad. Sci. ISSN 0077-8923

ANNALS OF THE NEW YORK ACADEMY OF SCIENCES
Issue: *Effects of Genome Structure and Sequence on Variation and Evolution*

Implications of genetic heterogeneity in cancer

Michael W. Schmitt, Marc J. Prindle, and Lawrence A. Loeb

Joseph Gottstein Memorial Cancer Research Laboratory, Department of Pathology, University of Washington, Seattle, Washington

Address for correspondence: Lawrence A. Loeb, 1959 NE Pacific St. Box 357705, Seattle, WA 98195. laloeb@uw.edu

DNA sequencing studies have established that many cancers contain tens of thousands of clonal mutations throughout their genomes, which is difficult to reconcile with the very low rate of mutation in normal human cells. This observation provides strong evidence for the mutator phenotype hypothesis, which proposes that a genome-wide elevation in the spontaneous mutation rate is an early step in carcinogenesis. An elevated mutation rate implies that cancers undergo continuous evolution, generating multiple subpopulations of cells that differ from one another in DNA sequence. The extensive heterogeneity in DNA sequence and continual tumor evolution that would occur in the context of a mutator phenotype have important implications for cancer diagnosis and therapy.

Keywords: mutator phenotype; cancer genomics; DNA sequencing; mutagenesis; tumor heterogeneity

Introduction

The evolution of a normal cell into a cancerous cell, with the capacity for sustained autonomous replication, implicates mutations in multiple genetic pathways that normally regulate growth, such as cell cycle control, metabolic pathways, and control of growth arrest and apoptosis.[1] That multiple mutations are required for a cell to gain the capability for uncontrolled replication is in apparent contrast to the high accuracy of DNA replication, which has been estimated to generate fewer than one mutation per 10 billion nucleotides replicated per cell division.[2] Owing to this discrepancy between the many mutations required for cancer and the extremely low rate at which mutations accumulate in normal human cells, we have long postulated that an early step in cancer evolution is an elevation in the spontaneous mutation rate, a concept known as the *mutator phenotype hypothesis*.[3] We have proposed that random genome-wide DNA-damaging events result in mutations in genes involved in maintaining genetic stability, leading to a cascading increase of mutations as a tumor evolves and progresses.[4]

The mutator phenotype hypothesis

The mutator phenotype hypothesis predicts that cancer development is essentially a random process driven by an elevated rate of accumulation of mutations throughout the genome. When proposed, this concept was in contrast to the prevailing idea that cancers have only a few recurrent mutations in key genes that would provide a limited number of targets for chemotherapy. DNA sequencing studies have now established that there is no single mutation that is invariably present in any specific type of cancer, nor is there a common consensus of mutations.[5] It is now clear that most human cancers contain many thousands of mutations throughout their genomes.[5] Yet even this large number is likely to be an underestimate, as prevailing approaches to DNA sequencing can detect only clonal mutations—that is, mutations that are present in a substantial fraction of the many millions of cells that constitute a detectable human cancer. Thus, it is likely that there are far more mutations scattered subclonally in only one or a few cells within a tumor that would be missed by current sequencing methods. Figure 1 emphasizes the genetic diversity that would arise as a consequence of a mutator phenotype. This total burden of clonal and low-abundance mutations in a tumor represents an enormous reservoir of genetic diversity that the tumor can draw upon to adapt; for example, it is conceivable that many tumors contain such a high mutational load that even before they are diagnosed tumor subclones are already

doi: 10.1111/j.1749-6632.2012.06590.x

Ann. N.Y. Acad. Sci. 1267 (2012) 110–116 © 2012 New York Academy of Sciences.

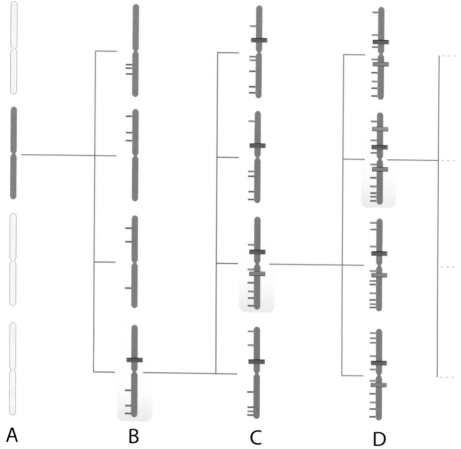

A   **B**   **C**   **D**

Figure 1. Continual mutagenesis results in subclonal genetic diversity. (A) A single cell (dark shading) acquires a mutator phenotype. (B) The mutator cell generates a series of progeny cells, each of which will possess a distinct set of random mutations (thin red lines). By chance, one of these mutations occurs in a gene that regulates growth (heavy red line), resulting in selective expansion of the cell containing this driver mutation. (C) The driver mutation, as well as all the random passenger mutations from the first round of mutagenesis, will be clonally present in the progeny cells. Ongoing mutagenesis generates an additional series of random mutations in each cell (thin green lines). Another driver mutation occurs (heavy green line), which will again result in selective expansion of a cell clone. (D) Ongoing rounds of random mutagenesis and selective expansion continue until sufficient driver mutations are present to result in uncontrolled growth and a clinically significant cancer. The driver and passenger mutations from the final clonal expansion will be present in the majority of the cells in the tumor. The mutant cells from earlier rounds of selection persist as subclonal populations, resulting in an enormous pool of genetic diversity within the tumor.

present that will prove to be resistant to targeted chemotherapeutics.

In consideration of a mutator phenotype in human cancer, we will first consider evidence from model systems that provide rationale for the concept. We will next examine data from human tumors that supply evidence for a mutator phenotype and will consider the apparent contrast between these data and the failure of sequencing studies to identify mutator mutations in human cancer. Finally,

we will discuss practical implications of a mutator phenotype in cancer.

A mutator phenotype can accelerate adaptation in *Escherichia coli*

While highly accurate DNA replication is critical for faithful transmission of the genome, occasional replication errors are essential to generate the genetic diversity that allows organisms to adapt to challenges in their environment. For example, when mutator bacteria (*E. coli* mutT1) are cocultured with wild-type strains under glucose-limiting conditions in a chemostat, the mutator strain outcompetes the wild-type bacteria,[6] presumably by generating diversity at a higher rate, thus enabling it to adapt more rapidly to glucose limitation and take over the culture. Similarly, mutator variants are seen to arise spontaneously in nature in response to stress.[7] *E. coli* undergoing the SOS response to DNA damage during replication display a short burst of mutation acquisition as error-prone polymerases are pressed into the task of copying past sites of DNA damage. This process not only allows the cells to proceed with replication in the presence of genotoxic stress, but also contributes to random, subclonal diversity that can lead to antibiotic resistance. Likewise, when a key DNA polymerase in *E. coli* (PolI) is randomly mutated to engineer a series of strains that encompass a broad range of mutation rates, cells expressing mutator alleles outcompete the wild-type cells when subjected to periodically fluctuating conditions of rapid growth followed by nutrient depletion and confluence.[8] Overall, these experiments establish that an elevated mutation rate provides a competitive growth advantage by facilitating accelerated adaptation to environmental challenges.

Evidence for a mutator phenotype in cancer

For a normal cell to transform into a cancer cell, it must overcome multiple mechanisms that normally regulate growth.[1] Then, as the cancer progresses, it continues to evolve, gaining the ability to invade, to metastasize, and to resist treatment. This continual refinement of fitness is analogous to that required in competition experiments in *E. coli*, and suggests that mutator cells would more rapidly generate the variation that would enable them to overcome limitations to growth than would cells that replicate their DNA with normal accuracy.

The conditions under which mutator cells possess a competitive advantage has been further probed by Beckman, who formulated a mathematical model to compare the probability of cancer arising with or without a mutator mutation.[9] In this model, progression to cancer is modeled as acquisition of 2–12 mutations of any of 100 oncogenes. Cell proliferation and death are assumed to be balanced until the critical number of oncogenic mutations are acquired, at which point selection results in expansion of the cell lineage. By modeling different rates of genome-wide random mutation, Beckman determined that mutator pathways are far more likely for most parameter values. Key parameters favoring mutator pathways include a larger number of mutations required to produce a cancer, and earlier onset of a cancer. Importantly, these conclusions are independent of absolute baseline mutation rates.

As this model predicts that rapidly mutating cells possess a competitive advantage over normal cells, it would be expected that alterations that elevate the basal mutation rate would result in enhanced tumorigenesis. Indeed, increased tumor incidence is observed both in mice that express mutated DNA polymerases that replicate DNA with altered accuracy in comparison with the wild-type allele,[10–12] and in mice with inactivation of specific steps in DNA repair pathways.[13] In addition, many inherited human mutations that inactivate DNA repair are associated with an increase in the incidence of multiple types of cancer.[14]

Mutations in human cancer

Recent evidence from the study of human cancers provides strong and complementary support for the concept of a mutator phenotype. Exons or entire genomes of human tumors have recently been sequenced.[16] These studies reveal that the cancer genome is punctuated with large numbers of single-base substitutions that are not present in normal DNA isolated from the same individual. In solid tumors, there are frequently 1,000 or more mutations detected in exons, with tens of thousands of total mutations per genome. In addition to single-base substitutions, cancers have been established to contain additional aberrations in DNA sequence, including insertions, deletions, rearrangements, and duplications. Larger scale alterations can be brought about by a variety of mechanisms, such as replication stress (see the review by Casper[17]), gene conversion, or through the poorly understood phenomenon of chromothripsis. Studies using current methods of DNA sequencing underestimate the true mutational load of each tumor, for they score predominantly for clonal mutations (i.e., those mutations present in a significant percentage of the tumor's cells). To further investigate the concept of a mutator phenotype in human cancer, we have developed a highly sensitive method, the random mutation capture assay, which can detect and identify nonclonal single-base substitutions in human cells.[15] Use of the assay in normal human tissues revealed that the frequency of unselected single base substitutions is $< 1 \times 10^{-8}$. By contrast, in six tumors that were analyzed, a mean of 210×10^{-8} mutations per base pair was detected, more than 200-fold greater than paired adjacent normal tissues.[18] This large increase in random, nonclonal mutations suggests that cancer cells replicate their DNA with greatly impaired accuracy relative to normal cells. Moreover, the emergence of resistance to chemotherapy is a prominent feature of cancer. There is evidence that chemotherapeutic-resistant mutant cells are present in a tumor as a subclonal population at the time of diagnosis, and that these mutant cells expand in response to the selective pressure of cancer therapy.[19,20] The preexistence of chemotherapy-resistant subclones further supports the concept that tumors encompass extensive genetic diversity far beyond that which can be detected by sequencing studies that primarily catalog clonal alterations. The tens of thousands of clonal mutations in tumors, which are likely to be accompanied by far more nonclonal mutations, are difficult to reconcile with the extremely high accuracy of DNA replication. This discrepancy suggests that impaired DNA replication fidelity may be a common feature of tumors.

The challenge of cataloguing the full scope of genetic variation in a tumor

Although the enormous burden of both clonal and subclonal mutations in human tumors is consistent with an elevated mutation rate as a common feature of cancer, DNA sequencing of human tumors has not revealed a substantial number of mutations in the genes that represent the major replicative DNA polymerases and repair proteins.[16] How can this disparity be reconciled? First, while the acquisition of mutations within these genes would result in an easily detected mutator phenotype, the efficacy and extent of DNA repair can be altered through a wide range of mechanisms,[21] including but not limited to DNA methylation, altered chromatin structure, mutations in nonprotein-coding regulatory regions, and altered activity of regulatory proteins. Indeed, altered expression of error-prone trans-lesion polymerases has been observed in multiple types of cancers; such altered expression may contribute to impaired replication of damaged DNA and muta-

genesis.[22–24] There is also emerging evidence that changes in the size or relative composition of nucleotide pools can lead to a mutator phenotype: in *E. coli*, efficient synthesis across replication-blocking DNA lesions requires elevated concentrations of nucleotides, and elevated nucleotide pools result in an increase in spontaneous mutagenesis *in vivo*.[25] Likewise, in *Saccharomyce cerevisiae*, genotoxic stress leads to dNTP pool expansion and an increase in spontaneous mutagenesis.[26]

The absence of obvious mutator mutations in human cancers may simply be a consequence of too narrow an understanding of which mutations might contribute to mutagenesis. It also is important to note that current methods of DNA sequencing of human tumors are best able to detect point mutations. Insertion and deletion events within genes are far more difficult to identify and cataloging this class of mutation requires sequencing to great depth.[27] Yet it is important to be aware of these events, as insertion or deletion of nucleotides within genes is more likely than point mutations to drastically alter protein function due to the effect of shifting the reading frame.[28]

An alternative explanation for the discrepancy between the many mutations in cancer and the general lack of mutator mutations identified in sequencing studies is that a mutator phenotype may be an early event in carcinogenesis that is subsequently selected against. In competition experiments performed in bacteria adapting to a changing environment, mutator variants initially out-compete the wild-type variants due to their greater capacity to evolve optimal fitness for their environment. However following extended passage for ~1,000 generations, the mutator variants acquire multiple deleterious substitutions and gradually lose fitness, resulting in a shift in equilibrium back toward bacterial strains with wild-type mutation rates.[29] We proposed that a similar process occurs in human cancers, such that an early event in carcinogenesis is acquisition of a mutator phenotype, followed by generation of genetic diversity and sequential rounds of selection for cell clones that possess mutations in genes that normally regulate cell growth. As a cancerous clone expands and becomes a clinically significant tumor consisting of 1×10^9 or more cells, a subpopulation that loses the mutator mutation (for example, by gene conversion of the mutant allele back to the wild-type allele[30] or by secondary somatic mutation[31]), yet retains the escape from cell growth restriction, would then have a selective advantage over clones that retain the mutator phenotype.

If a mutator mutation is indeed an early event in carcinogenesis, but is subsequently lost due to negative selection in the dominant clone(s) that expand to form the tumor, it would be missed by current DNA-sequencing methods, which can detect only predominant clones in a population. Mutator and other subclonal populations—which would be rare—could be detected only if tumor DNA is sequenced to exceptional depth. Detection of rare mutations is currently limited by the error rate of next-generation DNA sequencing, which ranges from 0.1–1%.[32] A search for subclonal mutator mutants will thus require the development of DNA sequencing methods with greatly increased accuracy.

Personalized therapy of evolving tumors

The mutator phenotype model predicts that cancers have a large pool of subclonal mutants, which is a concept with considerable therapeutic relevance. Indeed, mutants that confer resistance to targeted chemotherapeutic agents have been found to be present prior to the initiation of therapy.[19,20] Because preexisting mutant subclones are likely to exist, the mutator phenotype hypothesis predicts that single-agent cancer therapy frequently will have limited value. However, since the probability of having mutations that resist multiple therapies preexisting in a single clone is low, combined treatment with a number of agents would be more likely to result in effective control of cancer.

Another therapeutic approach to consider is based upon experiments in model systems that have established that an upper bound of genomic mutational burden exists.[33] Thus, if a mutator phenotype is indeed a common feature of human cancer, it may be possible to selectively target cancer cells through their elevated mutational burden. This upper bound has been exploited as a therapeutic target through an approach known as lethal mutagenesis: HIV is an organism known to have a high mutation rate,[34] and while this trait is beneficial in that the virus is able to rapidly evolve, further elevation of mutagenesis of HIV by treatment with mutagenic nucleoside analogs results in loss of fitness and extinction of HIV infection in cell culture.[35] Rationale for the same concept of lethal mutagenesis may also pertain to cancer cells. Inherited mutation in the BRCA1/2

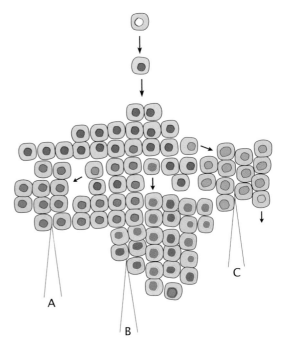

Figure 2. Spatial heterogeneity in cancer. Following clonal expansion of a cell that has acquired multiple driver mutations (red nucleus), mutagenesis persists resulting in accumulation of additional mutations and ongoing generation of subclonal populations within the tumor. Distinct subclones are depicted as cells with altered nuclei coloring. Spatial heterogeneity is an inherent property of a continually evolving tumor, and three biopsies of this same tumor (labeled as A, B, and C) will result in identification of three distinct DNA sequences. This topography indicates that multiple biopsies may be required to accurately assess the resistance of tumors to chemotherapy.

genes results in defective DNA repair, and further inhibition of DNA repair pathways with poly-ADP-ribose polymerase (PARP) inhibitors results in selective extinction of cancer cells.[36]

Since the mutator phenotype hypothesis predicts that tumors harbor a large number of subclones, DNA sequencing efforts that result in generation of a single DNA sequence from a heterogeneous population of millions of cells will not be representative of the overall genetic complexity of a tumor (Fig. 2). Sequencing tumor DNA to great depth and at high accuracy would give a better view of an individual tumor as a constantly evolving population of cells. Indeed, in studies that have sequenced DNA from both a primary tumor and a metastasis, there are mutations in the primary tumor that are not seen in the metastasis, suggesting that metastatic cells arise from tumor subclones.[37,38]

Greater depth of sequencing may allow for early detection of preexisting clones in a primary tumor with a "metastatic" genotype, suggesting that micrometastases are likely to already be present in the patient, a finding that would have important prognostic and therapeutic implications. Moreover early detection of drug-resistant subclones would allow for a more targeted form of chemotherapy, thus allowing for a longer period of response before drug resistance arises through further mutation or from the expansion of an even rarer subclone.

Ultimately, when cancer is viewed as a heterogeneous population of cells undergoing continual evolution in response to shifting selective pressures in an environment that includes host responses and cancer therapy, targeting that therapy based upon DNA sequences in a patient's tumor is likely to require multiple biopsies, with serial resequencing to identify the emergence of new dominant clones with differing patterns of chemotherapy sensitivity. Toward this end, new methods will need to be developed to sequence DNA from human tumors with increased accuracy and to significant depth. These methods are needed for a better understanding of genetic heterogeneity within human tumors, and to identify the contribution of a mutator phenotype to tumorigenesis and malignant progression.

Conflicts of interest

The authors declare no conflicts of interest.

References

1. Hanahan, D. & R.A. Weinberg. 2000. The hallmarks of cancer. *Cell* **100**: 57–70.
2. McCulloch, S.D. & T.A. Kunkel. 2008. The fidelity of DNA synthesis by eukaryotic replicative and translesion synthesis polymerases. *Cell Res.* **18**: 148–161.
3. Loeb, L.A., C.F. Springgate & N. Battula. 1974. Errors in DNA replication as a basis of malignant changes. *Cancer Res.* **34**: 2311–2321.
4. Salk, J.J., E.J. Fox & L.A. Loeb. 2010. Mutational heterogeneity in human cancers: origin and consequences. *Annu. Rev. Pathol.* **5**: 51–75.
5. Fox, E.J., J.J. Salk & L.A. Loeb. 2009. Cancer genome sequencing—an interim analysis. *Cancer Res.* **69**: 4948–4950.
6. Gibson, T.C., M.L. Scheppe & E.C. Cox. 1970. Fitness of an Escherichia coli mutator gene. *Science* **169**: 686–688.
7. Rosenberg, S.M. 2001. Evolving responsively: adaptive mutation. *Nat. Rev. Genet.* **2**: 504–515.
8. Loh, E., J.J. Salk & L.A. Loeb. 2010. Optimization of DNA polymerase mutation rates during bacterial evolution. *Proc. Natl. Acad. Sci. USA* **107**: 1154–1159.
9. Beckman, R.A. & L.A. Loeb. 2006. Efficiency of carcinogenesis with and without a mutator mutation. *Proc. Natl. Acad. Sci. USA* **103**: 14140–14145.
10. Albertson, T.M., M. Ogawa, J.M. Bugni, *et al.* 2009. DNA polymerase epsilon and delta proofreading suppress discrete mutator and cancer phenotypes in mice. 2009. *Proc. Natl. Acad. Sci. USA* **106**: 17101–17104.
11. Venkatesan, R.N., P.M. Treuting, E.D. Fuller, *et al.* 2007. Mutation at the polymerase active site of mouse DNA polymerase delta increases genomic instability and accelerates tumorigenesis. *Mol. Cell Biol.* **27**: 7669–7682.
12. Schmitt, M.W., R.N. Venkatesan, M.J. Pillaire, *et al.* 2010. Active site mutations in mammalian DNA polymerase delta alter accuracy and replication fork progression. *J. Biol. Chem.* **285**: 32264–32272.
13. Friedberg, E.C. & L.B. Meira. 2006. Database of mouse strains carrying targeted mutations in genes affecting biological responses to DNA damage Version 7. *DNA Repair* **5**: 189–209.
14. Preston, B.D., T.M. Albertson & A.J. Herr. 2010. DNA replication fidelity and cancer. *Semin. Cancer Biol.* **20**: 281–293.
15. Bielas, J.H. & L.A. Loeb. 2005. Quantification of random genomic mutations. *Nat. Methods* **2**: 285–290.
16. Loeb, L.A. 2011. Human cancers express mutator phenotypes: origin, consequences and targeting. *Nat. Rev. Cancer* **11**: 1–8.
17. Casper, A.M., D.M. Rosen & K.D. Rajula. 2012. Sites of genetic instability in mitosis and cancer. *Ann. N.Y. Acad. Sci.* **1267**: 24–30. This volume.
18. Bielas, J.H., K.R. Loeb, B.P. Rubin, *et al.* 2006. Human cancers express a mutator phenotype. *Proc. Natl. Acad. Sci. USA* **103**: 18238–18242.
19. Molinari, F., L. Felicioni, M. Buscarino, *et al.* 2011. Increased detection sensitivity for KRAS mutations enhances the prediction of anti-EGFR monoclonal antibody resistance in metastatic colorectal cancer. *Clin. Cancer Res.* **17**: 4901–4914.
20. Roche-Lestienne, C., J.L. Lai, S. Darre, *et al.* 2002. Several types of mutations of the Abl gene can be found in chronic myeloid leukemia patients resistant to STI571, and they can pre-exist to the onset of treatment. *Blood* **100**: 1014–1018.
21. Caporale, L.H. 2003. Natural selection and the emergence of a mutation phenotype: an update of the evolutionary synthesis considering mechanisms that affect genome variation. *Annu. Rev. Microbiol.* **57**: 467–485.
22. Pillaire, M.J., J. Selves, K. Gordien, *et al.* 2009. A "DNA replication" signature of progression and negative outcome in colorectal cancer. *Oncogene* **29**: 876–887.
23. Pan, Q., Y. Fang, Y. Xu, *et al.* 2005. Down-regulation of DNA polymerases κ, η, ι, and ζ in human lung, stomach, and colorectal cancers. *Cancer Lett.* **217**: 139–147.
24. Witkin, E.M. 1976. Ultraviolet mutagenesis and inducible DNA repair in *Escherichia coli*. *Bacteriol. Rev.* **40**: 869–907.
25. Gon, S., R. Napolitano, W. Rocha, *et al.* 2011. Increase in dNTP pool size during the DNA damage response plays a key role in spontaneous and induced-mutagenesis in Escherichia coli. *Proc. Natl. Acad. Sci. USA* **108**: 19311–19316.
26. Davidson, M.B., Y. Katou, A. Keszthelyi, *et al.* 2012. Endogenous DNA replication stress results in expansion of dNTP pools and a mutator phenotype. *EMBO J.* **31**: 895–907.

27. Walsh, T., S. Casadei, M.K. Lee, *et al.* 2011. Mutations in 12 genes for inherited ovarian, fallopian tube, and peritoneal carcinoma identified by massively parallel sequencing. *Proc. Natl. Acad. Sci. USA* **108:** 18032–18037.

28. Guo, H.H., J. Choe & L.A. Loeb. 2004. Protein tolerance to random amino acid change. *Proc. Natl. Acad. Sci. USA* **101:** 9205–9210.

29. Funchain, P., A. Yeung, J.L. Stewart, *et al.* 2000. The consequences of growth of a mutator strain of Escherichia coli as measured by loss of function among multiple gene targets and loss of fitness. *Genetics* **154:** 959–970.

30. Andersen, C.L., C. Wiuf, M. Kruhoffer, *et al.* 2007. Frequent occurrence of uniparental disomy in colorectal cancer. *Carcinogenesis* **28:** 38–48.

31. Norquist, B., K.A. Wurz, C.C. Pennil, *et al.* 2011. Secondary somatic mutations restoring BRCA1/2 predict chemotherapy resistance in hereditary ovarian carcinomas. *J. Clin. Oncol.* **29:** 3008–3015.

32. Flaherty, P., G. Natsoulis, O. Muralidharan, *et al.* 2011. Ultrasensitive detection of rare mutations using next-generation targeted resequencing. *Nucleic Acids Res.* **40:** 1–12.

33. Herr, A.J., M. Ogawa, N.A. Lawrence, *et al.* 2011. Mutator suppression and Escape from replication error–induced extinction in yeast. *PLoS Genet.* **7:** 1–16.

34. Preston, B.D., B.J. Poiesz & L.A. Loeb. 1988. Fidelity of HIV-1 reverse transcriptase. *Science* **242:** 1168–1171.

35. Loeb, L.A., J.M. Essigmann, F. Kazazi, *et al.* 1999. Lethal mutagenesis of HIV with mutagenic nucleoside analogs. *Proc. Natl. Acad. Sci. USA* **96:** 1492–1497.

36. Alli, E., V.B. Sharma, P. Sunderesakumar & J.M. Ford. 2009. Defective repair of oxidative DNA damage in triple-negative breast cancer confers sensitivity to inhibition of poly(ADP-ribose) polymerase. *Cancer Res.* **69:** 3589–3596.

37. Turajlic, S., S.J. Furney, M.B. Lambros, *et al.* 2011. Whole genome sequencing of matched primary and metastatic acral melanomas. *Genome Res.* Published in advance on December 19, 2011. doi:10.1101/gr.125591.111.

38. Vermaat, J.S., I.J. Nijman, M.J. Koudijs, *et al.* 2011. Primary colorectal cancers and their subsequent hepatic metastases are genetically different: implications for selection of patients for targeted treatment. *Clin. Cancer Res.* **18:** 688–699.

Ann. N.Y. Acad. Sci. ISSN 0077-8923

ANNALS OF THE NEW YORK ACADEMY OF SCIENCES

Issue: *Effects of Genome Structure and Sequence on Variation and Evolution*

Corrigendum for Ann. N.Y. Acad. Sci. 2009. 1182: 47–57

Tarhini, A.A. & J.M. Kirkwood. 2009. Clinical and immunologic basis of interferon therapy in melanoma. *Ann. N.Y. Acad. Sci.* **1182:** 47–57.

The following acknowledgment was inadvertently omitted from the above-cited article.

The project described was supported by Award Number P50CA121973 from the National Cancer Institute. The content is solely the responsibility of the authors and does not necessarily represent the official views of the National Cancer Institute or the National Institutes of Health.

doi: 10.1111/j.1749-6632.2012.06763.x